Alle Rechte, einschließlich das der Übersetzung und der Veranstaltung einer fremdsprachlichen Ausgabe, sind den Inhabern der Verlagsrechte vorbehalten.

Nachdruck und fotomechanische Wiedergabe, auch von einzelnen Seiten dieses Werkes, ohne Genehmigung verboten.

Printed in Germany · Imprimé en Allemagne

1991

ISBN 3-87 480-076-8

EUGEN G. LEUZE VERLAG · D-7968 SAULGAU/WÜRTT.

Satz: Ellen Wittel, D-7413 Gomaringen

Titel der dänischen Originalausgabe:
Specifikation for Flerlagsprint
Copyright © 1989 by Teknisk Forlag A/S, Copenhagen
ISBN 87-571-1175-8

Ausgearbeitet von der Gruppe der dänischen Leiterplattenverarbeiter	Prepared by the Danish PCB User's Group
Redigiert von Preben Lund Radiometer A/S	Edited by Preben Lund Radiometer A/S

Diese PERFAG-Spezifikation beschreibt die Qualitätsmerkmale von Mehrlagenschaltungen so, daß sie als Basis für eine Vereinbarung zwischen dem Kunden und dem Leiterplattenproduzenten dienen kann.

Wenn ein Leiterplattenhersteller die Fertigungsunterlagen von seinen Kunden erhält, so ist er verpflichtet, diese zu prüfen und Einspruch zu erheben, falls er seiner Erfahrung nach die Spezifikation nicht erfüllen kann. Außerdem muß der Hersteller reklamieren, wenn die Qualität der Filmvorlage die Erzeugung einer akzeptablen Leiterplattenqualität nicht zuläßt.

Wenn die PERFAG-Spezifikation und die Leiterplattenfertigungsunterlagen widersprüchlich sind, dann gelten immer die Fertigungsunterlagen. Übergeordnet gelten Kundenauftrag und Bestellbedingungen.

Die Vereinigung der dänischen Elektronikhersteller empfiehlt sowohl den Leiterplattenproduzenten als auch den Geräteherstellern innerhalb der Vereinigung die Anwendung der PERFAG-Spezifikationen.

The PERFAG specification lays down the quality level of multilayer boards so that it forms the basis of an agreement between the customer and the PCB manufacturer.

When receiving the PCB documentation, the PCB manufacturer is obliged to audit the documentation and object to such specifications which, according to his experience, cannot be met. Furthermore, the PCB manufacturer is obliged to object if the quality of the artwork is inadequate to ensure the desired longterm quality level.

In case of inconsistency between the PERFAG specification and the PCB documentation, the latter shall always be valid. The purchase order and its conditions, however, shall always take precedence.

The Association of Electronics Manufacturers in Denmark has recommended its members within the PCB Manufacturers' Group as well as the Instrument Manufacturer's Group, to use the PERFAG specifications.

Inhaltsverzeichnis	Seite / Page	Contents
Anwendungsbereich	7	Scope
Gültigkeit	7	Range of Validity
Ergänzende PERFAG-Spezifikation	7	Supplementary PCB Specifications
Literaturhinweise	7	References
1. Basismaterial	8	**1. Base Material**
1.1 Laminattyp	8	1.1 Type of Laminate
1.2 Dicke und Toleranzen von fertigen Leiterplatten	8	1.2 Thickness and Tolerance of Finished Boards
1.3 Dicke der Kupferfolie	8	1.3 Thickness of Copper Foil
1.4 Aufbau	9	1.4 Build-Up
1.5 Anforderungen an das Laminat fertiger Leiterplatten	11	1.5 Laminate Requirements of the Finished Boards
1.5.1 Allgemeine Anforderungen	11	1.5.1 General Requirements
1.5.2 Fleckenbildung (Measling)	11	1.5.2 Measling
1.5.3 Gewebezerrüttung (Crazing)	12	1.5.3 Crazing
1.5.4 Blasenbildung (Blistering)	12	1.5.4 Blistering
1.5.5 Delaminierung	13	1.5.5 Delamination
1.5.6 Hofbildung (Haloing)	13	1.5.6 Haloing
1.5.7 Gewebe-Strukturbildung (Weave Texture)	14	1.5.7 Weave Texture
1.5.8 Gewebefreilegung (Weave Exposure)	14	1.5.8 Weave Exposure
1.5.9 Unvollständige Aushärtung	14	1.5.9 Incomplete Curing
1.5.10 Metallische Einschlüsse	15	1.5.10 Metallic Inclusions
2. Galvanisierung	15	**2. Plating**
2.1 Allgemeine Qualitätsanforderungen	15	2.1 General Requirements
2.2 Galvanische Verkupferung	15	2.2 Copper Plating
2.3 Zinn/Blei-Galvanisierung	16	2.3 Tin/Lead Plating
2.4 Vergoldung	16	2.4 Gold Plating
3. Aufschmelzen galvanischer Zinn/Bleischichten auf durchkontaktierten Leiterplatten	17	**3. Reflowing of PTH Boards**
3.1 Allgemeine Qualitätsanforderungen	17	3.1 General Requirements
3.2 Ausführung	18	3.2 Execution
3.3 Bedeckung der Bohrungskante	18	3.3 Coverage of Hole-to-Pad Interface
4. Leiterbild	18	**4. Pattern**
4.1 Dokumentation des Leiterbildes	18	4.1 Documentation of Pattern
4.2 Allgemeine Anforderungen an das Leiterbild (75 %-Regel)	19	4.2 General Requirements of Pattern (75 % Rule)
4.3 Generelle Veränderungen des Leiterbildes	23	4.3 General Change of Pattern
4.4 Kantenschärfe des Leiterbildes	23	4.4 Edge Definition of Pattern
4.5 Einbuchtungen und Vorsprünge	24	4.5 Indentations and Projections
4.6 Fehlstellen und Nadellöcher	25	4.6 Voids and Pinholes
4.7 Metallpartikel	25	4.7 Metal Particles

	Inhaltsverzeichnis			Contents
4.8	Haftfestigkeit	25	4.8	Adhesion of Pattern
4.9	Abheben des Kupfers	26	4.9	Lifting of Copper
4.10	Lage des Leiterbildes auf SMT-Leiterplatten	26	4.10	Pattern Position on SMT Boards
4.11	Registrierung von Innenlagen	27	4.11	Registration of Inner Layers
4.12	Automatische SMD-Bestückung	28	4.12	Automatic Assembly of SMT Boards
4.13	Prüfung der fertigen Leiterplatte	30	4.13	Electrical Testing of Finished Boards
5.	**Metallisierte Löcher**	30	**5.**	**Plated-Through Holes**
5.1	Generelle Anforderungen	30	5.1	General Requirements
5.2	Dicke des galvanischen Niederschlages	31	5.2	Plating Thickness
5.3	Durchmessertoleranz	31	5.3	Tolerance of Diameter
5.4	Restringbreite	31	5.4	Annular Ring of the Solder Pad
5.5	Fehlstellen in metallisierten Löchern	35	5.5	Voids in Plated-Through Holes
5.6	Löt- und Entlötbeständigkeit	35	5.6	Soldering/Unsoldering Strength
5.7	Prüfabschnitt	36	5.7	Test Coupon
5.8	Epoxid-Smear (Verschmierung)	36	5.8	Epoxy Smear
5.9	Rückätzung des Basismaterials	37	5.9	Etchback of Base Material
5.10	Rückätzen des Kupfers	37	5.10	Etchback of Copper
5.11	Unebene Lochwand	38	5.11	Uneven Hole Wall
5.12	Porosität in der gebohrten Lochwand	38	5.12	Porosity in Drilled Hole Wall
5.13	Plating Pockets (Taschen in der Galvanikschicht)	38	5.13	Plating Pockets
5.14	Knospen	39	5.14	Nodules
5.15	Nagelkopfbildung	39	5.15	Nailheading
5.16	Grat	40	5.16	Burrs
5.17	Ablösung der Hülse vom Laminat	40	5.17	Laminate - Hole Wall Separation
5.18	Risse in der Metallisierung	41	5.18	Cracks in the Plating
5.19	Verbindung zwischen stromlos und galvanisch abgeschiedenem Kupfer	41	5.19	Plating Contact
5.20	Delaminierung	43	5.20	Delamination
6.	**Nicht metallisierte Löcher**	43	**6.**	**Nonplated-Through Holes**
6.1	Allgemeine Qualitätsanforderungen	45	6.1	General Requirements
6.2	Durchmessertoleranzen	44	6.2	Tolerance of Diameter
6.3	Restring	44	6.3	Annular Ring
7.	**Vergoldete Kontakte**	45	**7.**	**Gold Plated Contacts**
7.1	Allgemeine Qualitätsanforderungen	45	7.1	General Requirements
7.2	Galvanisierung	46	7.2	Plating
7.3	Nadellöcher (Pinholes)	46	7.3	Pinholes
7.4	Porosität	47	7.4	Porosity
7.5	Haftfestigkeit der galvanischen Beschichtung	47	7.5	Plating Adhesion
7.6	Ausrichtung von Ober- zu Unterseite	47	7.6	Side-to-Side Register
8.	**Siebgedruckte Masken und Komponentenbezeichnungen**	47	**8.**	**Screen-Printed Masks/Components Notations**
8.1	Definition	47	8.1	Definition
8.2	Einsatzbereich	48	8.2	Extent

8.3	Allgemeine Anforderungen	48	8.3	General Requirements	
8.4	Materialien	49	8.4	Materials	
8.5	Ausführung	49	8.5	Execution	
8.6	Überdrucken von Lötflächen (Pads)	50	8.6	Overprinting of Solder Pads	
8.7	Bedeckung von Leitern und Ausfüllung von Zwischenräumen	51	8.7	Coverage of Conductors and Filling of Conductor Spaces	
9.	**Fotopolymermasken**	52	**9.**	**Photopolymer Masks**	
9.1	Definition	52	9.1	Definition	
9.2	Einsatzbereich	52	9.2	Extent	
9.3	Allgemeine Qualitätsanforderungen	53	9.3	General Requirements	
9.4	Materialien	54	9.4	Materials	
9.5	Maskendicke und Leiterabstand	54	9.5	Thickness of Mask and Conductor Spacing	
9.6	Überdeckung von Lötflächen (Pads)	55	9.6	Overlapping of Solder Pads	
9.7	Leiterabdeckung	56	9.7	Coverage of Conductors	
10.	**Kohlepastendruck**	56	**10.**	**Carbon Ink Printing**	
10.1	Anwendung	56	10.1	Application	
10.2	Vorbedingung	56	10.2	Preconditions	
10.3	Allgemeine Anforderungen	56	10.3	General Requirements	
10.4	Materialien	57	10.4	Materials	
10.5	Detaillierungsgrad des Leiterbildes beim Siebdruck mit Kohlepaste	57	10.5	Degree of Carbon Pattern Detail	
10.6	Allgemeine Anforderungen an das Leiterbild (75 %-Regel)	58	10.6	General Requirements of Pattern (75 % Rule)	
10.7	Kantenschärfe des Kohleleiterbildes	59	10.7	Edge Definition of Carbon Pattern	
10.8	Einbuchtungen und Vorsprünge	59	10.8	Indentations and Projections	
10.9	Kohlerückstände	60	10.9	Carbon Specks	
10.10	Fehlstellen in der Kohleschicht	60	10.10	Voids in the Carbon Layer	
10.11	Kohleüberlappung	61	10.11	Carbon Overlapping	
10.12	Anschlußflächen	61	10.12	Termination Area	
10.13	Registrierung des Kohleleiterbildes	63	10.13	Registration of Carbon Pattern	
10.14	Haftfestigkeit	63	10.14	Adhesion	
10.15	Widerstand und Isolationswiderstand	63	10.15	Resistance/Insulation Resistance	
10.16	Oberflächenschutz (abziehbare Maske)	65	10.16	Surface Protection (Peelable Mask)	
10.17	Ausführung	66	10.17	Execution	
10.18	Leiter unter der Kohleschicht	66	10.18	Conductors under Carbon	
10.19	Auswirkung des Lötprozesses	67	10.19	Effect of Soldering	
10.20	Prüfung unter Umweltbedingungen	67	10.20	Environmental Testing	
11.	**Abziehbare Masken**	67	**11.**	**Peelable Masks**	
11.1	Anwendungsbereich	67	11.1	Application	
11.2	Ausführung	67	11.2	Execution	
11.3	Allgemeine Anforderungen	67	11.3	General Requirements	
11.4	Besondere Anforderungen	68	11.4	Special Requirements	
12.	**Oberflächenschutz**	70	**12.**	**Protective Coating**	
12.1	Allgemeine Anforderungen	70	12.1	General Requirements	

	Inhaltsverzeichnis			Contents
12.2	Zinn/Blei-Galvanisierung mit nachfolgendem Aufschmelzen	70	12.2	Tin/Lead Plating with Reflowing
12.3	Heißluftverzinnung (Hot Air Levelling)	70	12.3	Solder Coating and Hot-Air Levelling
12.4	Schutzlackierung	71	12.4	Protective Lacquering
12.5	Chemische Verzinnung	71	12.5	Immersion Tin
12.6	Andere Schutzbeschichtungen	71	12.6	Other Types of Protective Coating
13.	**Löten**	71	**13.**	**Soldering**
13.1	Allgemeine Anforderungen	71	13.1	General Requirements
13.2	Beanspruchung beim Löten	72	13.2	Effect of Soldering
13.3	Lötbarkeit nach Lagerung	73	13.3	Solderability after Storage
14.	**Mechanische Bearbeitung**	73	**14.**	**Machining**
14.1	Allgemeine Anforderungen	73	14.1	General Requirements
14.2	Wölbung und Verwindung	73	14.2	Warp and Twist
14.2.1	Definition (für rechteckige Leiterplatten)	73	14.2.1	Definitions (Rectangular Boards)
14.2.2	Bestimmung von Wölbung und Verwindung	74	14.2.2	Determination of Warp and Twist
14.2.3	Normal zulässige Werte für Wölbung und Verwindung	75	14.2.3	Normal Warp and Twist Requirements
14.2.4	Schärfere Anforderungen für Wölbung und Verwindung	76	14.2.4	Tighter Warp and Twist Requirements
14.3	Referenzsystem	77	14.3	Reference System
14.4	Bemaßung des Leiterplattenumrisses	78	14.4	Dimension of the Board Contour
14.5	Bemaßung von im Nutzen angeordneten Leiterplatten	80	14.5	Dimensioning of Panelized Boards
14.6	Festlegung des Umrisses	82	14.6	Determination of Contour
14.7	Bearbeitungstoleranzen	83	14.7	Machining Tolerances
14.8	Umrißbearbeitung	83	14.8	Machining the Contour
14.9	Ritzen	83	14.9	Scoring
14.10	Kodierschlitze in Randkontaktleisten	84	14.10	Edge Connector Polarization Slot
14.10.1	Lage des Schlitzes	84	14.10.1	Position of the Slot
14.10.2	Herstellung des Kodierschlitzes	85	14.10.2	Machining of the Slot
14.11	Anfasung des Randkontaktes	85	14.11	Chamfering of Edge Connector
14.12	Locharten	86	14.12	Types of Holes
14.12.1	Bohrungen für die manuelle Bestückung	86	14.12.1	Holes for Manual Assembly
14.12.2	Bohrungen für die automatische Bestückung	86	14.12.2	Holes for Automatic Assembly
14.12.3	Aufnahmebohrungen für die automatische Bestückung	86	14.12.3	Tooling Holes for Automatic Assembly
14.12.4	Montagebohrungen	87	14.12.4	Mounting Holes
14.13	Festlegung der Lochposition	87	14.13	Determination of Hole Position
14.14	Lagetoleranzen für Bohrungen	87	14.14	Positional Tolerances on Holes
	Stichwortverzeichnis, deutsch	89		**Index, german**
	Stichwortverzeichnis, englisch	97		**Index, english**

Anwendungsbereich

Diese Spezifikation gilt für Mehrlagen-Leiterplatten (Multilayer), und zwar sowohl für die Lochmontage- als auch für die Oberflächenmontagetechnik (SMT). Unter der Bezeichnung "Pad" sind ebenso Lötaugen mit Bohrungen für die Lochmontage wie Anschlußflächen (Footprints) für SMT zu verstehen.

Abkürzungen:
HMT = Lochmontagetechnik
SMT = Oberflächenmontagetechnik

Gültigkeit

Zur Spezifikation neuer Leiterplatten ist nur die Ausgabe B der PERFAG 3-Spezifikation gültig.

Bereits bestehende Leiterplattenspezifikationen, die aufgrund einer früheren PERFAG-Ausgabe erstellt wurden, bleiben jedoch weiterhin gültig, auch wenn die ältere PERFAG-Spezifikation stellenweise von der Ausgabe B abweicht.

Ergänzende PERFAG-Spezifikationen

PERFAG 10: "PCB Manufacturing Documentation Data, Format, Tables und Media", veröffentlicht Ende 1989.

PERFAG 11: "Measuring Methods: Test of Printed Circuit Boards" veröffentlicht Ende 1989.

Literaturhinweise

IPC-A-600	Acceptability of Printed Boards
IPC-TM-650	Test Methods Manual
IPS-S-804	Solderability Test Methods
IEC-68-2-30	Basic Environmental Testing Procedures

Quality Assessment of Printed Circuit Boards*
(Bishops Graphics, Inc., 1985)

* Beziehbar über: Data Elektronik Konsult AB, P.O. Box 1321
S-430 30 Fnllesås, Schweden

Scope

This specification is valid for multilayer boards intended for conventional leaded components (for hole mounting) and/or leadless SMT components (for surface mount technology). By solder pads is understood conventional HMT solder pads (with solder holes) as well as SMT solder pads (footprints) for surface mount technology (without solder holes).

Abbreviations:
HMT = Hole Mount Technology
SMT = Surface Mount Technology

Range of Validity

Only issue B of PERFAG 3 is valid for specifying new PCBs to a PCB manufacturer.

If, in the case of PCBs specified according to previous PERFAG issues, specifications occur that cannot be met due to changed specifications in the new PERFAG issue, the previous PERFAG issues are valid for the specifications in question.

Supplementary PCB Specifications

PERFAG 10: "PCB Manufacturing Documentation Data, Format, Tables and Media" To be published at the end of 1989

PERFAG 11: "Measuring Methods: Test of Printed Circuit Boards" Published at the end of 1989

References

IPC-A-600	Acceptability of Printed Boards
IPC-TM-650	Test Methods Manual
IPS-S-804	Solderability Test Methods
IEC 68-2-30	Basic Environmental Testing Procedures

Quality Assessment of Printed Circuit Boards*
(Bishop Graphics, Inc., 1985)

* Available from: Data Elektronik Konsult AB, P.O. Box 1321
S-430 30 Fnllesås, Sweden

| Basismaterial | Base Material |

1. Basismaterial | 1. Base Material

1.1 Laminattyp | 1.1 Type of Laminate

Starres Basismaterial:	FR4
Prepreg (B-Zustand):	FR4

Der Laminattyp wird normalerweise in der Spezifikation angegeben.

Rigid material:	FR4
Prepreg (B-stage):	FR4

The type of laminate will usually be stated on the master drawing.

1.2 Dicke und Toleranz von fertigen Leiterplatten | 1.2 Thickness and Tolerance of Finished Boards

Übliche Leiterplattendicken sind

0,8; 1,0; 1,2; 1,6; 2,4/2,5 und 3,2 mm

Empfohlen werden Dicken von 1,6, 2,4 und 3,2 mm. Die Dickentoleranz beträgt ± 10 %.

Die Dicke wird wie folgt gemessen:
Vom Leiterplattenhersteller: Direkt nach dem Verpressen, d.h. einschließlich der Kupferkaschierung der Außenlagen.
Vom Kunden: Vorzugsweise an den gedruckten Randkontakten, falls vorhanden, sonst an einer freigeätzten Stelle, wobei in diesem Fall die Kupferschichtdicke hinzugerechnet und die Dicke einer eventuell vorhandenen Maske abgezogen werden muß.

Designhinweis
Bei steckbaren Leiterplatten ist auf ein ausreichendes Spiel zwischen Leiterplatte und Führung zu achten. Gegebenenfalls ist die Lötstoppmaske zurückzusetzen.

Normally occurring thickness are

0.8; 1.0; 1.2; 1.6; 2.4/2.5 and 3.2 mm

Recommended thickness are 1.6, 2.4 and 3.2 mm. The thickness tolerance is ± 10 %.

The thickness is to be measured as follows:
By the PCB manufacturer: Immediately after lamination, i.e., incl. the copper foil of the outer layers.

By the customer: Preferably across the edge connector contacts, if any, otherwise across a clean-etched laminate with the addition of nominal thickness of the outer layers' copper foil and deduction of nominal thickness of solder masks, if any.

Design Note
In the case of plugable boards, a suitable clearance in the board guides with respect to the total board thickness should be ensured, possibly by withdrawing solder masks, if any, from the board edges.

1.3 Dicke der Kupferfolie | 1.3 Thickness of Copper Foil

a. **Innenlagen**

Normalausführung	35 µm
Bearbeitet	min. 27 µm
Spezialausführung	70 µm
Bearbeitet	min. 56 µm

Bemerkung
Falls notwendig, kann die Schichtdicke durch galvanisches Verstärken einer dünneren Kupferfolie, z.B. in Verbindung mit buried via holes (vergrabenen Durchgangsbohrungen) erreicht werden.

a. **Inner Layers:**

Normal execution:	35 µm
Finished thickness:	min. 27 µm
Special execution:	70 µm
Finished thickness:	min. 56 µm

Note
If convenient, the thickness can be achieved by copper plating on top of a thinner copper foil, e.g., in conjunction with buried via holes.

Basismaterial / Base Material

b. **Außenlagen**

5; 9; 17,5; 35 und 70 μm

Falls nicht in der Spezifikation vermerkt, kann die Dicke der Kupferfolie vom Leiterplattenhersteller entsprechend Punkt 4.2 gewählt werden. Mindestdicke von Kupferfolie + galvanische Kupferschicht: 30 μm.

Bemerkung
Eine reine Additivtechnik darf nur nach vorheriger Absprache mit dem Kunden angewendet werden.

b. **Outer Layers:**

5; 9; 17.5; 35 and 70 μm

If not stated on the master drawing, the thickness of the copper foil can be chosen by the PCB manufacturer in accordance with Item 4.2. Minimum thickness of copper foil + copper plating: 30 μm.

Note
Additive techniques are not allowed unless previously agreed upon.

1.4 Aufbau / Build-Up

a. Ein nicht spezifizierter Aufbau sollte symmetrisch in Bezug auf die mittlere Lage erfolgen.

Designhinweis
Es wird eine gerade Lagenzahl empfohlen.

b. Um einen korrekten Aufbau sicherzustellen, werden die einzelnen Lagen mit 1, 2, 3, ... oder A, B, C, ... gekennzeichnet. Es ist empfehlenswert, die Kennzeichnung so anzubringen, daß sie von der Oberseite aus lesbar ist (s. Skizze: 1 oder A).

a. Unspecified build-up should be symmetrical with respect to the centre plane.

Design Note
An even number of layers is recommended.

b. The layers are marked 1, 2, 3, ... or A, B, C, ..., to ensure a correct build-up. It is recommended that the marking be direct-reading when seen from the upper side (1 or A).

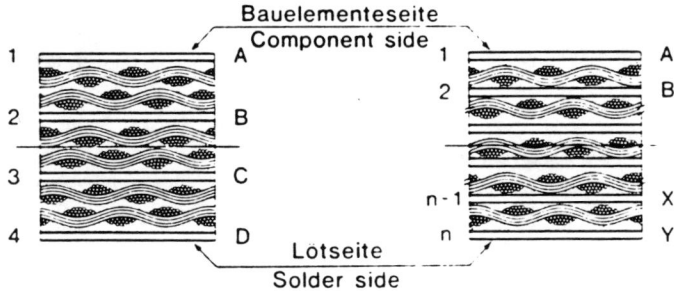

Designhinweis
Der korrekte Aufbau läßt sich leicht kontrollieren, wenn am Rand jeder Lage ein kleiner Kupferstreifen vorgesehen wird, der von Lage zu Lage versetzt ist und beim Übereinanderschichten ein treppenförmiges Bild ergibt.

Design Note
An easy check of correct build-up can be achieved by placing a small copper area on each layer along one of the board edges. If the copper areas are regularly displaced with respect to each other, a correct build-up shows up as a regular staircase when viewing the edge.

Basismaterial / Base Material

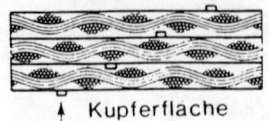
Kupferflache / Copper area

Befinden sich auf einer oder mehreren Lagen Masse- oder Versorgungsebenen, dann sollten diese etwas von den oben erwähnten Kupferflächen zurückgesetzt sein.

If the layers contain ground or voltage planes, such planes should be withdrawn a little around the copper area mentioned above.

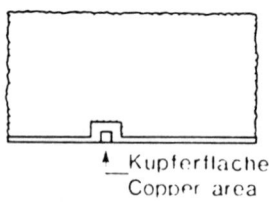
Kupferflache / Copper area

c. Normalerweise bleibt dem Leiterplattenhersteller der Aufbau eines Multilayers im Detail überlassen, wenn es um die Anzahl und die Dicke von Prepregs, die Lagenanordnung auf Kernen aus starrem Basismaterial und die Entscheidung, ob für die Außenlagen Kupferfolie plus Prepregs oder kupferkaschiertes starres Laminat verwendet wird, geht. Siehe Punkt e und f hinsichtlich besonderer Anforderungen bezüglich der Dicke und Dickentoleranzen des dielektrischen Materials.

c. Normally it is left to the PCB manufacturer to determine the detailed build-up regarding the number and thickness of the prepregs, the distribution of the layers on inner cores of rigid material, including the decision to use copper foil + prepregs or rigid copper-clad material for the outer layers. See Items e and f regarding special requirements for the thickness and thickness tolerance of the dielectric material.

d. Die Dicke der Kupferfolie ist für die einzelnen Lagen anzugeben.
Designhinweis
Starre innenliegende Kerne sollten auf beiden Seiten mit Kupferfolien gleicher Dicke kaschiert sein.

d. The copper foil thickness of the individual layers must be specified.
Design Note
Inner cores of rigid base material should have the same thickness of copper foil on both sides.

e. Nur wenn die Dicke T_p des dielektrischen Materials zwischen den einzelnen Ebenen von Bedeutung ist, z.B. wegen der Impedanz, sollte diese vorgeschrieben werden. Die Dicke T_p wird von Kupfer zu Kupfer oder von Galvanikschicht zu Galvanikschicht gemessen, vgl. Punkt 2.1 e.

e. Only where the thickness T_p of the dielectric material between the individual layers is important, e.g., regarding the impedance, should the thickness be stated. The thickness T_p is measured from copper to copper, or from plating to plating, cf. Item 2.1 e.

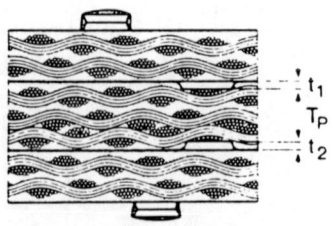

Prepreg Layers
Any thickness T_p can be achieved by a suitable combination of prepreg thicknesses, possibly combined with rigid base material without copper foil. The absolute minimum value of T_p after lamination is 75 µm.

Rigid Base Material
It is recommended that the PCB manufacturer be consulted regarding thicknesses.

f. Only where the thickness tolerance of the dielectric material between the individual layers is important, should the tolerance ΔT_p be stated. Normative values versus the thickness T_p of the dielectric material are stated below:

Nominal thickness T_p	Tolerance ΔT_p
$T \leq 0.125$ mm	± 0.025 mm
$0.150 \leq T \leq 0.175$ mm	± 0.038 mm
$0.200 \leq T \leq 0.300$ mm	± 0.050 mm
$0.360 \leq T \leq 0.510$ mm	± 0.064 mm
$0.560 \leq T \leq 0.760$ mm	± 0.076 mm

1.5 Laminate Requirements of the Finished Boards

1.5.1 General Requirements

The board must be clean with no dirt, dust or drill chips etc. on the surface or in the holes

In Items 1.5.2 through 1.5.10 below some general flaws in the laminate of the finished board are elaborated.

1.5.2 Measling

Measling appears as white spots or crosses just below the board surface and is caused by small cavities occurring where the fibres of the glass weave intersect.

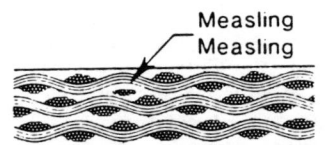

Basismaterial	Base Material

a. Measlingflecke sind dann zulässig, wenn sie weit verstreut mit einer durchschnittlichen Dichte von maximal 50 Flecken pro 100 cm² vorkommen.

b. Die Measlingflecken dürfen keine freiliegenden Glasfasern aufweisen.

c. Es dürfen sich mehrere Measlingflecken nicht berühren.

d. Measlingflecke, die gleichzeitig zwei benachbarte Leiterzüge oder Lötaugen berühren, sind nicht zulässig.

1.5.3 Gewebezerrüttung (Crazing)

Crazing, das auf oder unter der Leiterplattenoberfläche vorkommen kann, läßt sich am besten als Verbindung mehrerer Measlingflecken beschreiben. Crazing tritt häufig auf mechanisch bearbeiteten Oberflächen oder um größer gebohrte oder gestanzte Löcher herum auf.

a. Crazing ist ein Rückweisungsgrund.

---

a. Measle spots are tolerable if widely scattered with an average density of max. 50 spots per 100 cm².

b. No glass fibres shall be exposed in the measle spots.

c. No measle spots shall touch each other.

d. Solder pads and conductors shall not be connected by measle spots.

1.5.3 Crazing

Crazing, which can occur on or below the board surface, is best described as merging measle spots. Crazing occurs frequently on machined surfaces and around larger drilled or punched holes.

a. Crazing is cause for rejection.

Gewebezerruttung / Crazing

b. Crazing kann dann als Ausnahme akzeptiert werden, wenn der betroffene Bereich mehr als 10 mm Abstand zu Leiterzügen oder Lötaugen hat.

1.5.4 Blasenbildung (Blistering)

Als Blasenbildung wird ein lokales, mit Delaminierung verbundenes Aufschwellen des Basismaterials bezeichnet. Es kann sowohl zwischen den Glasfasergewebelagen als auch zwischen Basismaterial und Kupfer auftreten. Siehe auch Punkt 13.2 b.

---

b. Exemption can be granted when the area exhibiting crazing is located more than 10 mm away from solder pads or conductors.

1.5.4 Blistering

Blistering is characterized as a local swelling and delamination of the base material, either between some of the glass cloth layers or between the base material and the copper foil. See also Item 13.2 b.

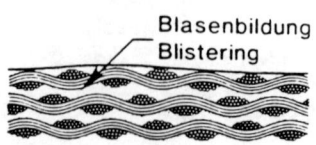
Blasenbildung / Blistering

Basismaterial	Base Material

a. Blasenbildung ist ein Rückweisungsgrund.

b. Eine Ausnahme kann zugestanden werden, wenn die Leiterplatte von Hand gelötet werden soll, und wenn maximal zwei Blasen auf 100 cm^2 vorkommen. Wenn das der Fall ist, darf die Blasenfläche 1 cm^2 nicht überschreiten.

1.5.5 Delaminierung

Im Gegensatz zur Blasenbildung (Blistering) ist die Delaminierung eine großflächige Trennung einzelner Glasfasergewebematten voneinander oder eine Ablösung der Kupferfolie vom Basismaterial. Siehe auch Punkt 5.20 und 13.2 b.

a. Blistering is cause for rejection.

b. Exemption can be granted if the board is to be hand-soldered, provided no more than two blisters per 100 cm^2 are found. If so, a blister should not be larger than 1 cm^2.

1.5.5 Delamination

Contrary to blistering, delamination is a widespread separation between the individual glass cloth layers of the base material and the copper foil. See also Items 5.20 and 13.2 b.

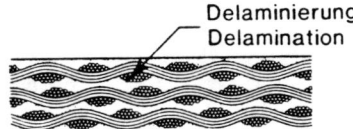
Delaminierung / Delamination

a. Delaminierung ist ein Rückweisungsgrund.

1.5.6 Hofbildung (Haloing)

Haloing tritt als helle Fläche auf bearbeiteten Bereichen, zum Beispiel als heller Ring um eine Bohrung, in Erscheinung und zwar entweder direkt auf der Oberfläche oder innerhalb des Basismaterials.

a. Delamination is cause for rejection.

1.5.6 Haloing

Haloing appears as a light area in the machined areas of the board, e.g., as a light ring around a hole, either directly on the surface or in the base material.

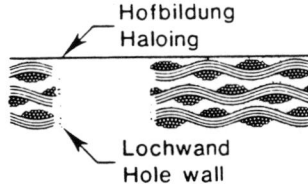
Hofbildung / Haloing
Lochwand / Hole wall

a. Haloing zwischen zwei Leitern oder Lötaugen ist ein Rückweisungsgrund.

b. Eine Ausnahme kann gemacht werden, wenn das Haloing vollständig getrennt von den Leiterzügen auftritt.

a. Haloing occurring between two conductors or solder pads is cause for rejection.

b. Exemption can be granted when the haloing occurs completely isolated from the circuits.

Basismaterial / Base Material

1.5.7 Gewebestrukturbildung (Weave Texture)

Das Glasgewebe wird sichtbar, wenn die äußere Epoxidschicht zwar das Gewebe noch verdeckt, jedoch sehr dünn ist.

1.5.7 Weave Texture

Visible glass weave occurs where the outermost epoxy layer is very thin but still covering the glass weave.

Gewebestrukturbildung / Weave texture

a. Dies ist zulässig, wenn keine Glasfasern gebrochen sind, und wenn das Glasfasergewebe vollständig mit Epoxid bedeckt ist.

a. Weave texture is tolerable if no glass fibres are broken and the glass weave is fully covered with epoxy.

1.5.8 Gewebefreilegung (Weave Exposure)

Freiliegendes Gewebe tritt dann auf, wenn auf der Substratoberfläche die Epoxidschicht ganz oder stellenweise fehlt.

1.5.8 Weave Exposure

Uncovered glass weave on the board surface is due to the lack of an epoxy layer at the surface.

Gewebefreilegung / Weave exposure

a. Gewebefreilegung ist ein Rückweisungsgrund.

a. Weave exposure is cause for rejection.

1.5.9 Unvollständige Aushärtung

Eine unvollständige Aushärtung des Epoxids im Basismaterial zeigt sich dann, daß die Substratoberfläche etwas weniger hart als normal ist. Eine ungenügende Aushärtung kann beim Maschinenlöten zum Auftreten von Netzen (Lotfäden) führen und ist daher nicht akzeptabel.

Außerdem kann die Netzbildung auch durch falsche Lötparameter, wie z.B. durch unkorrektes Fluxen, hervorgerufen werden. Der Anwender sollte daher die Ursachen ergründen, z.B. durch das Löten von Leiterplatten aus anderen Fertigungschargen.

1.5.9 Incomplete Curing

Incomplete curing of the board's epoxy is detected by the surfaces being a little less hard than normal. Insufficient curing can result in webbing during mass soldering, which is unacceptable.

Webbing, however, can also be caused by inadequate soldering parameters, e.g. incorrect fluxing. The user should therefore identify the reason, e.g. by soldering boards from other deliveries.

a. Unvollständige Aushärtung ist ein Rückweisungsgrund.

a. Incomplete curing is cause for rejection.

1.5.10 Metallic Inclusions

Metallic particles encapsulated in the base material are acceptable under the following conditions:

a. They are smaller than 1 mm in any direction.
b. They are at least 0.5 mm away from any active part of the circuit.
c. There are max. 2 particles per 100 cm^2.

2. Plating

2.1 General Requirements

a. No area shall exhibit "burning".
b. The plating shall adhere well to the underlay. see Item 7.5.
c. A few spurious deposits occurring at random shall be removed without any essential deterioration of appearance or insulation resistance. See Items 4.7, 8.3 f and 9.3 f.
d. Minor scratching of the tin/lead will be accepted provided the underlying copper is not exposed.

2.2 Copper Plating

a. **Holes**

   Copper thickness* in every
   single hole   min. 25 μm

   Minimum thickness within small
   local areas of the hole wall   15 μm

   Voids in the hole wall   See Item 5.5

* Determined as the average value of 6 measurements, 3 on each side of a microsection, at the following depths: about 1/4, 1/2 and 3/4 down. Isolated thin or thick areas shall be avoided by displacing the measuring point.

Note 1
The values imply that $d \geq 0.25\, t$ where t indicates the thickness of the board, and d indicates the diameter of the hole.

Galvanisierung / Plating

Bemerkung 2
Für 0,2 t ≤ d < 0,25 t ändern sich die Werte 25 und 15 μm auf 18 bzw. 12 μm.

Note 2
For 0.2 t ≤ d < 0.25 t, the values 25 and 15 μm are changed to 18 and 12 μm, respectively.

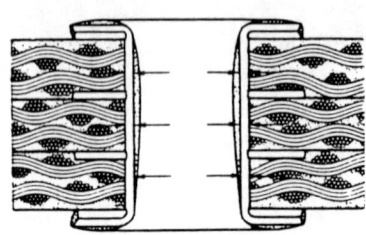

b. **Leiterbild**

Mindestschichtdicke des galvanischen Kupferauftrages:	25 μm
Mindestschichtdicke des galvanischen Kupfers plus Kupferfolie:	30 μm

b. **Pattern on Outer Layers**

Minimum thickness of copper plating	25 μm
Minimum thickness of copper plating + copper foil	30 μm

2.3 Zinn/Blei-Galvanisierung

a. Legierungszusammensetzung, Nominalwert: 60/40

b. Zinngehalt:
zulässige Streuung über die Leiterplattenfläche min. 55 %
max. 70 %

Hinweis
Wenn in der Leiterplattenspezifikation angegeben, kann die Zinn/Bleibeschichtung einschließlich Aufschmelzen durch eine Heißluftverzinnung (Hot Air Levelling) ersetzt werden (vergl. Punkt 12.3).

2.3 Tin/Lead Plating

a. Composition of alloy, nominal value 60/40

b. Content of tin:
Accepted variation across finished board min. 55 %
max. 70 %

Note
When stated on the master drawing, the tin/lead plating, including reflowing, can be replaced with solder coating and hot-air levelling, cf. Item 12.3.

2.4 Vergoldung

Die Spezifikation gilt ebenso für Rand-Steckkontakte wie für selektiv vergoldete Flächen wie Schalterkontakte und Tastaturkontakte.

a. **Unterschicht**

Nickel	spannungsarm
Reinheit	min. 99,5 %
Schichtdicke	min. 5 μm
	max. 15 μm

2.4 Gold Plating

The specification is also valid for edge connectors and selectively gold plated areas, e.g., switch contacts and keyboard switches.

a. **Underlay**

Nickel	Low stress
Purity	min. 99.5 %
Thickness of layer	min. 5 μm
	max. 15 μm

b.	**Deckschicht**		b.	**Top Layer**	
	Gold	Hartgold, legiert		Gold	Alloy-hardened gold
	Schichtdicke*	min. 1,5 μm		Thickness of layer*	min. 1.5 μm
	Härte (HV)	1,4 - 2 kN/mm²		Hardness (HV)	1.4 - 2 kN/mm²
	Reinheit	min. 99,7 %		Purity	min. 99.7 %

    * mit einer radiographischen Methode gemessen.

    * To be measured by a radiographic method.

c. **Allgemeine Qualitätsanforderungen** siehe Abschnitt 7.

c. **General Requirements**   See Item 7

3. Aufschmelzen galvanischer Zinn/-Blei-Schichten auf durchkontaktierten Leiterplatten

3. Reflowing of PTH Boards

3.1 Allgemeine Qualitätsanforderungen

a. Nach dem Aufschmelzen dürfen keine Zinn/Blei-Überhänge mehr vorhanden sein. Das Zinn/Blei muß auf Pads (Lötaugen und -flächen) und Leitern vollständig geschmolzen sein. Bei großen Masseebenen oder ähnlichem können Ausnahmen gemacht werden. Die Leiterflanken müssen nicht bedeckt sein.

b. Nach dem Aufschmelzen dürfen keine nicht benetzten Stellen auftreten. Pads dürfen keine Entnetzung aufweisen, während eine Entnetzung auf 25 % der Leiter- oder Masseebenenfläche zugestanden werden kann.

c. Die Zinn/Blei-Oberfläche muß auf allen Pads eben und glänzend sein.

d. Zulässige Oberflächenfehler, angegeben in Prozent einer Leiterfläche oder Masseebene.

    körnige (sandige) Oberfläche   5 %
    graue Oberfläche   10 %

Anmerkung
Diese Fehler dürfen nicht auf Lötaugen vorkommen.

e. Erstarrungslinien (Krokodilhautmuster) gelten nicht als Fehler.

f. Es darf keine Verfärbung des Basismaterials auftreten.

3.1 General Requirements

a. Any tin/lead overhang shall be removed by reflowing. The tin/lead shall have been completely molten on pads, in holes and on conductors. Exemption for incomplete melting of large ground planes or the like can be granted. Coverage of conductor edges is not required.

b. After reflowing, no areas exhibiting nonwetting shall be found. Dewetting shall not occur on solder pads, whereas it is accepted that 25 % of the area of a conductor or of a ground plane can show dewetting.

c. The tin/lead surface shall be even and bright on all solder pads and in all hole barrels.

d. Accepted surface defects, indicated in percentage of the area of a conductor or a ground plane:

    gritty surface   5 %
    grey surface   10 %

Note
These defects shall not occur on solder pads.

e. Freeze lines (crocodile pattern) are not regarded as a defect.

f. Discolouration of the base material shall not occur.

3.2 Ausführung

Alle mit Zinn/Blei galvanisierten Leiterplatten müssen aufgeschmolzen werden, sofern nichts anderes vorgeschrieben ist.

3.3 Bedeckung der Bohrungskante

Nach dem Aufschmelzen soll der Übergang zwischen Bohrung und Lötauge glatt sein. Die Dicke t der Zinn/Blei-Schicht in der Übergangszone muß genügen, um nach einer Lagerung der Leiterplatten die Lötbarkeit zuverlässig sicherzustellen, vgl. Abschnitt 13.3.

3.2 Execution

All tin/lead plated boards shall be reflowed unless otherwise indicated.

3.3 Coverage of Hole-to-Pad Interface

After reflowing, the hole-to-pad interface shall be smooth, and the thickness t of the tin/lead layer at the interface zone shall be sufficient to ensure full solderability after storage of the boards, cf. Item 13.3.

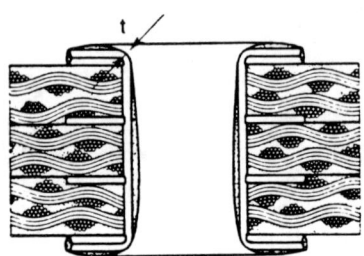

4. Leiterbild

4.1 Dokumentation des Leiterbildes

Das Leiterbild ist in den Fertigungsunterlagen und -daten des Kunden dokumentiert. Die Dokumentation kann in verschiedenen Formen vorliegen:

a) als geplotteter Film aus CAD-Daten,
b) als fotografisch aufgenommener Film von einer manuell angefertigten Vorlage,
c) als CAD-Plotterdaten entsprechend PERFAG 10 oder einer besonderen Vereinbarung.

Wenn der Kunde Filme entsprechend a) oder b) zur Verfügung stellt, dann dienen diese Filme auch als Referenz zur Beurteilung des Leiterbildes der fertigen Leiterplatte (relative Lage, Leiterbreiten, Lötflächenabmessungen, usw.). Der Kunde kann identische Filme zurückbehalten und bei der Wareneingangskontrolle verwenden.

4. Pattern

4.1 Documentation of Pattern

The pattern is determined by the customer's documentation which can be available in several ways:

a) as plotted films derived from CAD plotter data
b) as films photographically reproduced from manual artwork
c) as CAD plotter data according to PERFAG 10 or to a previous arrangement.

When the customer delivers filmwork according to items a and b above, this filmwork serves as a reference to the pattern on the outer and inner layers of the finished board (relative placement, conductor widths, solder pad dimensions, etc.). The customer may retain identical filmwork for use during incoming inspection of the finished boards.

Stellt der Kunde CAD-Plotterdaten entsprechend Punkt c zur Verfügung, dann sind zusätzliche Angaben über Mindestbreite von Leiterzügen und Isolationsabstände sowie Bohr- und Umrißzeichnungen erforderlich. Wenn keine anderen Vereinbarungen getroffen werden, ist der Leiterplattenproduzent voll für die Herstellung der Leiterplatten genau nach den angelieferten CAD-Daten verantwortlich.

Die angelieferten CAD-Daten werden zur Herstellung von Fotowerkzeugen verwendet, die jedoch aus produktionstechnischen Gründen modifiziert werden können. Das setzt aber voraus, daß die fertigen Leiterplatten innerhalb der zulässigen Toleranzen den angelieferten CAD-Daten entsprechen. Das bedeutet auch, daß im Leiterbild weder Teile entfernt noch hinzugefügt werden dürfen, z.B die Einführung von Ausgleichsflächen. Nach Absprache soll der Leiterplattenproduzent dem Kunden Filme zur Verfügung stellen, die genau den vorgegebenen CAD-Daten entsprechen.

Designhinweis

Es wird empfohlen, das Leiterplattendesign so anzulegen, daß über das gesamte Leiterbild in etwa die gleiche Schaltungsdichte erzeugt wird. Das gilt sowohl für jede Seite an sich als auch für eine Seite im Vergleich zur anderen. Dies bietet wesentliche Vorteile bezüglich einer gleichmäßigeren galvanischen Abscheidung und führt zu geringerer Wölbung und Verwindung.

When the customer delivers CAD plotter data according to Item c above, information on minimum values of conductor widths and insulation distances as well as on the drill and contour drawings should also be given. Unless other arrangements are made, the PCB manufacturer has the full responsibility for manufacturing the boards in agreement with the CAD data delivered to him.

The CAD plotter data delivered form the basis of generating the phototools which may be modified to compensate for pattern changes during manufacture. This implies that the finished boards conform to the original CAD plotter data within the tolerances given. It also implies that no part be removed from or added to the pattern, e.g., the introduction of dummy areas. By agreement, the PCB manufacturer shall deliver a set of films unmodified with respect to the original CAD plotter data and graphically identical to the data.

Design Note

It is recommended that the PCB designer himself perform a balancing of the pattern to achieve approximately the same pattern density all over the board, partly on the individual side, partly between the two sides. This implies substantial advantages regarding a more uniform plating of the board and also a reduced amount of warp and twist.

4.2 Allgemeine Anforderungen an das Leiterbild (75 %-Regel)

Bei jeder zufällig vorkommenden Kombination der unter Punkt 4.3 bis 4.7 angegebenen Anforderungen bezüglich des Leiterbildes sind die im folgenden unter a bis c angeführten primären Forderungen einzuhalten. Alle Messungen gelten für die vertikale Projektion P auf die Leiterplatte.

4.2 General Requirements on Pattern (75 % Rule)

With any arbitrary combination of the requirements stated in Items 4.3 to 4.7 concerning the pattern, the following primary requirements a to c shall be fulfilled. All measurements are based on the vertical projection P on the board.

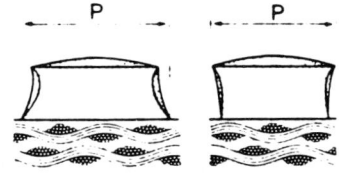

Pattern

a. The effective width of a conductor shall not be reduced below 75% of the nominal value*. This applies also to the conductor foot F regardless of the vertical projection. Therefore, a possible verification implies the preparation of microsections.

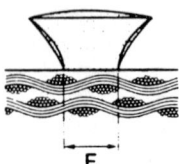

Design Note
The above requirements ensure the adhesion of the conductor to the laminate.

When a certain resistance is required, i.e., a minimum reduction of the conductor's sectional area, a note should be stated on the master drawing.

b. The diameter of an HMT solder pad or the width and length of an SMT solder pad shall not be reduced below 75% of the nominal values* because of locally occurring defects, e.g., indentations along the pad edge, cf. Item 4.5.

Note 1
Displacements of pattern edges beyond the values stated in Item 4.3 are not covered by Item b and therefore unacceptable.

Note 2
Unless an exemption has been granted, the requirements to the annular ring stated in Item 5.4 shall be fulfilled, particularly the size of the break-out of the annular ring as indicated in Item 5.4 d.

c. The insulation distance R between the individual parts of the pattern (conductors and solder pads, etc.) shall not be reduced below 75% of the nominal distance A*. Unwanted random metal particles shall be considered when assessing the reduction of the insulation resistance.

* Reference: Customer's documentation, cf. Item 4.1.

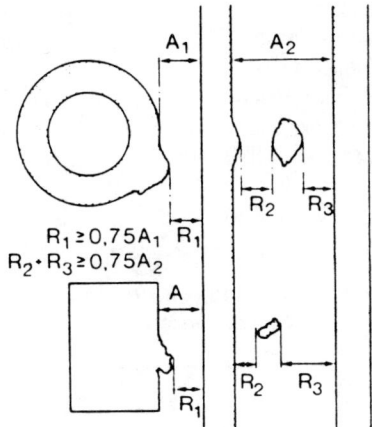

4.3 Generelle Veränderungen des Leiterbildes

Einzelne Bestandteile des Leiterbildes, z.B. ein Leiter, können aufgrund von Belichtungsfehlern, zu schwachem oder zu starkem Ätzen oder von Galvanisierfehlern (aber nicht wegen Registrierfehlern) breiter oder schmaler werden. Das bedeutet einen Kantenversatz nach innen oder außen.

Der Leiterplattenhersteller muß die Qualität der angelieferten Dokumentation sehr sorgfältig prüfen. Wenn die Dicke der Kupferfolie nicht auf der Zeichnung angegeben ist, ist sie entsprechend der Pakkungsdichte (Leiterbreite und -abstand) zu wählen. Der Leiterplattenhersteller muß immer die unter Punkt 2.2 b genannte Anforderung bezüglich der Gesamtdicke von galvanischem Kupferniederschlag plus Kupferfoliendicke erfüllen.

Bei mit Zinn/Blei beschichteten und aufgeschmolzenen Leiterplatten, bei heißluftverzinnten sowie Blankkupferleiterplatten soll der Einwärts- oder Auswärtsversatz ΔF^* (Breitenabweichung) des Leiterbildrandes gegenüber seiner nominalen Lage geringer als nachfolgend angegeben sein.

Dicke der Kupferfolie	Breitenabweichung ΔF
µm	µm
5	15
9	19
17,5	25
35	38
70	63

4.3 General Change of Pattern

The individual parts of the pattern, e.g., a conductor, may become a little wider or narrower as a result of imaging faults, too weak or too heavy etching, or plating growth, excluding misregistration. This means that the conductor edges are subject to displacements (inwards or outwards).

The PCB manufacturer must assess very carefully the quality of the documentation received. If the thickness of the copper foil has not been stated on the master drawing, it shall be chosen according to the pattern density (conductor width and spacing). The PCB manufacturer must always fulfil the requirements in Item 2.2 b regarding the total thickness of the copper plating and the copper foil.

In the case of tin/lead plated and reflowed boards, solder-coated and hot-air levelled boards and bare-copper boards, the inward or outward displacement ΔF^* of the pattern edge relative to its nominal location shall be less than stated below. The value applies also to inner layers whether these are plated or nonplated.

Thickness of Copper Foil	Displacement ΔF
µm	µm
5	15
9	19
17.5	25
35	38
70	63

| Leiterbild | Pattern |

Die Breitenabweichung ΔF wird als die halbe Breitenänderung der Leiterbreite gemessen. Dabei gilt der Originalfilm des Kunden als Referenz, vgl. Punkt 4.1. Die in den Punkten 4.4 und 4.5 angeführten Unregelmäßigkeiten werden bei der Bestimmung der Breitenabweichung nicht berücksichtigt.

The displacement ΔF is measured as half the change in conductor width, taking the customer's original film as a reference, cf. Item 4.1. The irregularities stated in Items 4.4 and 4.5 are excluded when determining the displacement.

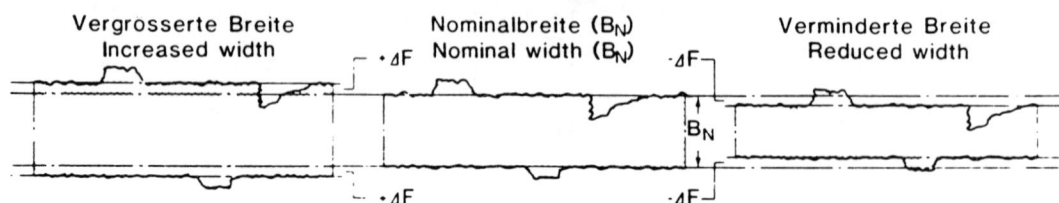

Bemerkung 1
Die 75 %-Regel (Punkt 4.2) schreibt geringere Abweichungen gegenüber den Nominalabmessungen des Leiterbildes vor:

Dicke der Kupferfolie	Minimale nominale Leiterbreite und minimaler Isolationsabstand
μm	mm
5	0,12**
9	0,15**
17,5	0,20**
35	0,30
70	0,50

** Dabei ist vorausgesetzt, daß das Leiterbild möglichst gleichmäßig dicht angelegt ist.

Bemerkung 2
Der Ätzfaktor a = t/s muß ≥ 1 sein.

Note 1
The 75 % rule (Item 4.2) implies a lower limitation of the nominal pattern dimensions:

Minimum Nominal Thickness of Copper Foil	Conductor Width and Insulation Distance
μm	mm
5	0.12**
9	0.15**
17.5	0.20**
35	0.30
70	0.50

** It is implied that the pattern be balanced to achieve a reasonably even pattern density

Note 2
The etch factor a = t/s must be ≥ 1

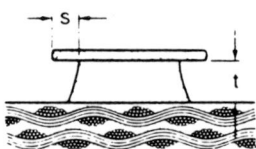

Bemerkung 3
Im allgemeinen kann das Leiterbild nur bei Leiterbreiten und Isolationsabständen ≥ 0,3 mm im Siebdruck aufgebracht werden.

Note 3
In general, image transfer by screen printing can be used only in the case of conductor widths and insulation distances ≥ 0.3 mm

Designhinweis

Wenn ein Leiter um einen Winkel θ > 90° umgelenkt wird, so sollte der Winkel abgerundet oder abgestumpft werden. Beim Fotoplotten läßt sich der Winkel bei A abrunden, bei B bleibt jedoch der spitze Winkel bestehen, wodurch die Gefahr auftritt,

- daß sich bei der Verwendung vonTrockenfilmlötstoppmasken Lufteinschlüsse bilden und
- daß im Siebdruck aufgebrachter Lötstopplack den Winkel mangelhaft ausfüllt.

Design Note

Where a conductor is bent and forms an angle θ > 90°, the angle should be rounded or truncated. Photoplotting will round the angle at A but retain the sharp angle at B where there is a risk of:

- air pockets in the case of a dry-film solder mask

- ink skipping in the case of a screen-printed solder mask

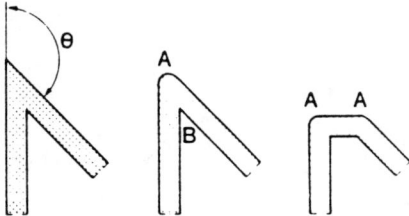

4.4 Kantenschärfe des Leiterbildes

Die Ungleichmäßigkeit (Kantenschärfe) U der Kanten des Schaltungsmusters, gemessen von der Spitze bis zum Grund, muß unter den nachfolgend angegebenen Werten liegen:

Sollmaß von Leiterbahnbreite und Isolationsabstand mm	Zulässige Ungleichmäßigkeit (Spitze bis Grund) mm
0,12	≤ 0,030
0,15	≤ 0,038
0,20	≤ 0,050
0,30	≤ 0,075
0,50	≤ 0,100

4.4 Edge Definition of Pattern

The unevenness (edge definition), U, crest to trough, of the pattern edges, shall be less than the values stated below:

Nominal Conductor Width and Insulation Distance mm	Acceptable Unevenness (Crest to Trough) mm
0.12	≤ 0.030
0.15	≤ 0.038
0.20	≤ 0.050
0.30	≤ 0.075
0.50	≤ 0.100

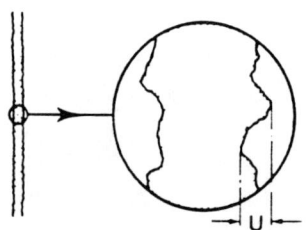

4.5	**Einbuchtungen und Vorsprünge**	4.5	**Indentations and Projections**

Einige wenige unerwünschte Einbuchtungen und Vorsprünge an den Leiterkanten sind unter folgenden Bedingungen zulässig:

a. Die Tiefe D einer Einbuchtung muß kleiner als 1 mm sein.
b. Die Höhe H eines Vorsprunges muß kleiner als 1 mm sein.
   Hinweis
   Die 75 %-Regel (Punkt 4.2) besagt, daß 1 mm große Einbuchtungen oder Vorsprünge nur akzeptiert werden können, wenn Leiterbreiten bzw. -abstände ≥ 4 mm sind.

A few unwanted indentations and projections along the pattern edges are accepted under the following conditions:

a. The depth D of an indentation shall be less than 1 mm.
b. The size H of a projection shall be less than 1 mm.
   Note
   The 75 % rule (Item 4.2) implies that indentations or projections of 1 mm are accepted only in cases of conductor widths, alternatively insulation distances, of 4 mm and above.

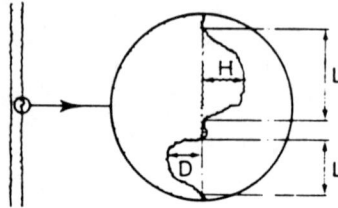

c. Die Länge L einer Einbuchtung oder eines Vorsprunges muß < 2 mm sein. Für Lötflächen (Pads) gelten folgende zusätzliche Einschränkungen:

   Lötaugen für die Lochmontage (HMT)
   Länge L ≤ 90° oder 25 % des Umfanges

   SMT-Lötflächen
   Länge L ≤ 25 % der Kantenlänge, pro Kante gemessen.

c. The length L of an indentation or a projection shall be less than 2 mm. The following additional limitations are valid for solder pads:

   HMT solder pads
   The length L ≤ 90° or 25 % of the circumference.

   SMT solder pads
   The length L ≤ 25 % of edge length, per edge.

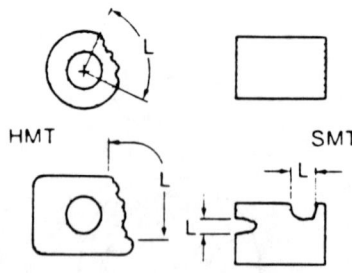

Leiterbild	Pattern

d. Es darf durchschnittlich weniger als 1 Fehler (Vorsprung oder Einbuchtung) auf 50 mm Kantenlänge auftreten.

d. The average number of defects (projections and indentations) along a pattern edge shall be less than one defect per 50 mm of edge length.

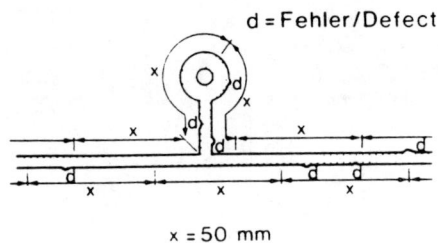

d = Fehler/Defect

x = 50 mm

4.6 Fehlstellen und Nadellöcher

Unerwünschte Fehlstellen und Nadellöcher im Schaltungsmuster sind unter folgenden Bedingungen zulässig:

a. Der Lochdurchmesser ist kleiner als 0,15 mm.

b. Die Lochanzahl muß im Durchschnitt geringer als 1 Loch pro 50 mm Leiterlänge sein.

c. In einem Pad (Lötauge oder SMD-Lötfläche) darf höchstens 1 Loch auftreten. Löcher im Übergang vom Pad zum Leiter sind unzulässig.

4.6 Voids and Pinholes

Unwanted voids and pinholes in the pattern are accepted under the following conditions:

a. The hole diameter is less than 0.15 mm.

b. The average number of holes in a conductor shall be less than one hole per 50 mm of conductor length.

c. No more than one hole per solder pad shall be found. The hole shall never occur at the pad-to-conductor interface.

4.7 Metallpartikel

Unerwünschte, fest am Basismaterial haftende Metallpartikel, die durch unvollständiges Ätzen verursacht werden, sind unter folgenden Bedingungen zulässig:

a. Die größte Abmessung ist 1 mm.

b. Der Mindestabstand zwischen den Partikeln beträgt 50 mm.

4.7 Metal Particles

Unintentional metal particles sticking to the base material, caused by incomplete etching, are accepted under the following conditions:

a. The largest dimension is 1 mm.

b. The distance between particles is greater than 50 mm.

4.8 Haftfestigkeit

a. **Leiter**
    Nach dem Löten entsprechend Abschnitt 13.2 soll die Haftung der Kupferfolie auf dem Basismaterial so stark sein, daß die Abschälkraft für über 0,8 mm breite Leiter folgende Werte übersteigt:
    1,4 N/mm für Epoxid-Glashartgewebe
    Der Test wird nach IPC-TM-650, Punkt 2.4.8 ausgeführt.

4.8 Adhesion of Pattern

a. **Conductors**
    After soldering according to Item 13.2, the adhesion of the copper foil to the base material shall be high enough to ensure that the peel strength, valid for conductors above 0.8 mm in width, is greater than:
    1.4 N/mm for epoxy glass
    The test is made in accordance with IPC-TM-650, Item 2.4.8.

b. **SMD-Anschlußflächen (Pads)**

SMD-Lötflächen müssen ein fünfmaliges Anlöten und Entlöten eines 0.5 mm dicken Drahtes aushalten und nach der vollständigen Abkühlung eine Abziehkraft von mindestens 5 N/mm^2 aufweisen. Die Lötfläche muß mindestens 1,5 x 2 mm groß sein. Die Temperatur der Lötkolbenspitze soll zwischen 232 und 260 °C liegen, und die Wärme soll über den Draht und nicht direkt der SMD-Lötfläche zugeführt werden.

b. **Solder Pads for Surface Mounting**

Solder pads shall withstand five times soldering and unsoldering a 0.5 mm wire, and after full cooling exhibit a pull-off strength of min. 5 N/mm^2. The test implies solder pad dimensions of min. 1.5 x 2 mm. The temperature of the tip of the soldering iron shall lie within the interval 232 - 260 °C, and the heat shall be transferred via the wire, not directly to the solder pad.

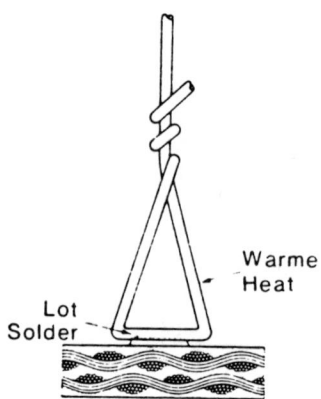

4.9 Abheben des Kupfers

Es darf sich kein Teil des Leiterbildes vom Basismaterial abheben.

4.9 Lifting of Copper

No part of the pattern shall be lifted from the base laminate.

4.10 Lage des Leiterbildes auf SMT-Leiterplatten

Um eine fehlerfreie SMD-Bestückung sicherzustellen, werden an die Lagegenauigkeit des Leiterbildes hohe Anforderungen gestellt. Das setzt eine gute Zusammenarbeit zwischen Leiterplattenhersteller und Leiterplattendesigner voraus.

Der Leiterplattenhersteller muß die ihm angelieferten Fertigungsunterlagen sorgfältig kontrollieren, besonders wenn sie von Hand erstellt werden. Außerdem muß er die Fertigungsprozesse so steuern, daß prozeßbedingte Abweichungen minimiert werden. Zum Beispiel kann es notwendig sein, zunächst das Basismaterial zu stabilisieren, um permanente Dimensionsänderungen zu reduzieren.

4.10 Pattern Position on SMT Boards

To ensure a flawless surface mounting of leadless components (SMDs) on the board, very high requirements are set to the correct position of the patterns on the finished board. Close collaboration between the PCB manufacturer and the PCB designer is therefore implied.

The PCB manufacturer must carefully evaluate the quality of the PCB documentation received, especially when it is generated manually. Furthermore, he must adapt the manufacture to minimize displacements caused by the processes. For example, it may be necessary to stabilize the base material prior to manufacture in order to reduce the permanent dimensional change.

Der Versatz ΔF des Leiterbildes kann folgendermaßen bestimmt werden:

a. Direktes Ausmessen der Leiterbildlage, wobei die Dokumentation des Kunden, d.h. die angelieferten Unterlagen, als Referenz dienen (vgl. Punkt 4.1).

b. Auflegen und Ausrichten des originalen Kundenfilmes und, falls nötig, Ausmessen der Verschiebungen.

In beiden Fällen, beim Messen bzw. beim Ausrichten, ist das Referenzsystem maßgebend. Vgl. Absatz 14.3, in dem das Referenzsystem beschrieben ist.

Es besteht die Forderung, daß der Versatz ΔF für Leiterplattenabmessungen bis 200 x 200 mm in jeder Richtung kleiner als 0,10 mm sein muß. Bei größeren Leiterplatten kann pro angefangene 100 mm (über 200 mm hinaus) 0,05 mm addiert werden. Das gilt sowohl für die Einzelplatten als auch für im Nutzen ausgelieferte Leiterplatten bis zu einer Größe von 450 x 450 mm.

The displacement ΔF of the pattern can be determined as stated below:

a. A direct measurement of the pattern position with respect to the customer's documentation, cf. Item 4.1.

b. Overlaying and aligning the customer's original film on the board with a subsequent measurement of displacements, if any.

In both cases, measurement and alignment, respectively, are based on the datum and axis directions of the reference system, cf. Item 14.3, where the reference system is defined.

It is required that the displacement ΔF in any direction be less than 0.10 mm for board sizes up to 200 x 200 mm. In case of larger boards, 0.05 mm is added for every commenced 100 mm of length beyond 200 mm. This applies to individual boards, alternatively boards delivered in panels, up to 450 x 450 mm.

4.11 Registrierung von Innenlagen

Während des Verpressens eines Multilayers können die Innenlagen etwas von der Sollposition weggleiten. Das Leiterbild (Leiter oder Masse-/Versorgungsebenen) kann dadurch soweit verrutschen, daß ein kritischer Versatz zwischen dem Leiterbild und der Lochmetallisierung, die in einem späteren Prozeß hergestellt wird, auftritt.

Damit auf den Innenlagen ein Isolationsabstand von mindestens 0,15 mm eingehalten wird, darf keine der Innenlagen im Vergleich zu einer anderen einen größeren Versatz ΔF als 0.30 mm aufweisen. Siehe Skizze. Außerdem ist es erforderlich, daß kein Innenlagenpad/keine Innenlagenaussparung mehr als 0,30 mm im Vergleich zum gebohrten Loch versetzt ist. Das gilt sowohl für einzelne als auch für im Nutzen gefertigte Leiterplatten bis zu einer Größe von 450 x 450 mm.

4.11 Registration of Inner Layers

During the lamination press cycle the inner layers can flow a little away from the nominal location. The pattern (conductor or ground/voltage planes) may hereby be displaced so much that a serious misregistration occurs between the pattern and the barrels of the plated-through holes made in the later processes.

In order to maintain an insulation distance of min 0.15 mm on the inner layers of the finished board, it is required that none of the inner layers exhibit a larger mutual displacement ΔF than 0.30 mm. See the figure below. Furthermore, it is required that no inner layer pad/inner layer aperture be displaced more than 0.30 mm relative to the drilled hole. This applies to individual boards, as well as boards delivered in panels, up to 450 x 450 mm.

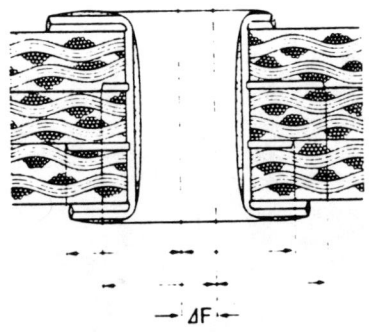

Design Note (cf. Design Note of Item 1.4)
A certain checking of correct registration of the layers can be achieved by placing a small copper area on each layer along each board edge. If the documentation exhibits a correct registration of the copper areas with respect to each other, a correct registration of the layers of the finished board will show up as "ladders" with no displaced steps when viewing the boards edges. Displacements in the central areas of the board can only be revealed by an X-ray examination or by preparing microsections.

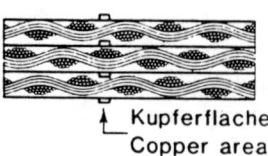
Kupferflache
Copper area

If the layers contain ground or voltage planes, such planes should be withdrawn a little around the copper areas mentioned above.

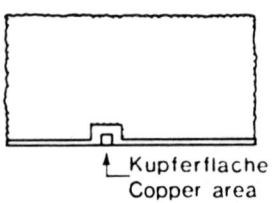
Kupferflache
Copper area

4.12 Automatic Assembly of SMT Boards

In particular, surface mounted boards invite automation of the assembly process, but this requires special measures when designing the boards. Cf. Design Note 3 of Item 14.5 concerning panelized boards.

a. **Edge Clearance Areas**

   Moving the board through the automatic assembly line requires a certain edge clearance area A outside the effective board area, so that the board can be moved along the guide rails of the line. The edge clearance area necessary

Die erforderlichen Freiflächen hängen von dem verwendeten Bestückungsautomaten ab, werden aber oft als 10 mm breite Streifen an zwei gegenüberliegenden Leiterplattenseiten vorgesehen. Werden Schlitze K vorgefräst, so lassen sich die Randstreifen nach dem Bestücken leicht abtrennen.

depends upon the automatic placement machine used, but it frequently occurs as 10 mm wide strips along two of the board edges. By appropriate contouring, K, the strips can be broken off after assembly has been completed.

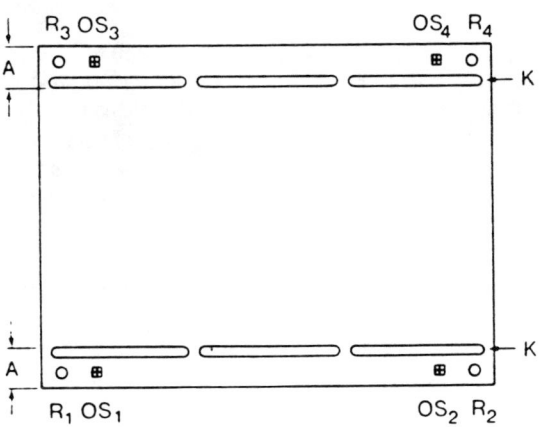

b. **Aufnahmebohrungen**

Zur Positionierung der Leiterplatte in dem Bestückungsautomaten können Aufnahmebohrungen R in die Randstreifen eingebracht werden. Vergleiche Designhinweis 3 in Absatz 14.3, ebenso Absatz 14.12.3. Die Anzahl der Aufnahmebohrungen kann variieren, z.B. R_1 und R_2, wobei R_2 als Langloch ausgeführt sein kann, um die Positionierung der Leiterplatte zu erleichtern. In anderen Fällen werden 3 oder 4 Aufnahmebohrungen verwendet (R_1, R_2, R_3 und eventuell R_4).

c. **Optische Registriermarken**

Wenn der Bestückungsautomat mit einem optischen Positioniersystem ausgerüstet ist, ist es zweckmäßig, in der Nähe der Aufnahmebohrungen optische Registriermarken (Fiducials) anzubringen. Diese können sich entweder auf den freien Randflächen oder auf der Leiterplatte selbst befinden. Es sollen mindestens 3 optische Registriermarken OS_1, OS_2, OS_3 verwendet werden. Sie dürfen nicht von einer Maske (Lötstopp- oder Isoliermaske) überdeckt werden. Die optischen Registriermarken werden geätzt und dienen als Bezugspunkte für das Leiterbild. Nachstehend sind Beispiele für optische Registriermarken dargestellt.

b. **Tooling Holes**

For the sake of positioning the board in the automatic placement machine, tooling holes R can be located in the edge clearance strips. Cf. Design Note 3 of Item 14.3. Cf. also Item 14.12.3. The number of tooling holes can vary, e.g., 2 tooling holes R_1 and R_2, of which R_2 can be made oblong to facilitate positioning the board. In other cases, 3 to 4 tooling holes are used (R_1, R_2, R_3 and possibly R_4).

c. **Optical Targets**

If the automatic placement machine is provided with an optical positioning system, it is expedient to include optical targets OS (fiducials), not too far away from the tooling holes R, either within the board area or outside in the edge clearance areas. At least 3 optical targets should be used, OS_1, OS_2 and OS_3, and they shall not be covered with a mask (solder or insulation mask). The optical targets are etched in the copper and used as reference for the pattern. Examples of optical targets are shown overleaf.

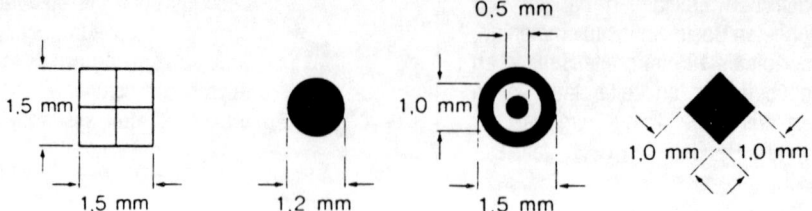

Designhinweis
Um Rückfragen des Leiterplattenherstellers zu vermeiden, sollte die Fertigungszeichnung des Kunden den Vermerk "optische Registriermarken für die Bestückung" sowie die Darstellung einer Registriermarke enthalten.

4.13 Prüfung der fertigen Leiterplatte

Die nachfolgend angeführten Tests werden nach Kundenanforderung, oder wenn sie prozeßbedingt erforderlich sind, durchgeführt.

a. Elektrische Prüfung auf Kurzschlüsse und Unterbrechungen.
b. Automatische optische Kontrolle (AOI: Automatic Optical Inspection).

Hinweis
Wenn die Leiterplatte mit Masken beschichtet werden soll, muß die Kontrolle eventuell vor dem Aufbringen der Masken erfolgen.

5. Metallisierte Löcher

5.1 Generelle Anforderungen

a. Die Löcher im Basismaterial sollen so gebohrt sein, daß die Anforderungen beim Einlöten und an die Auslötbeständigkeit (siehe Punkt 5.6) erfüllt werden, d.h. daß die Lochwände eine gewisse Rauhigkeit aufweisen müssen. Gleichzeitig sollen gute Voraussetzungen für das Löten geschaffen werden, siehe Punkt 13.2. D.h., daß die Lochwände eine so geringe Unebenheit aufweisen sollen, daß eine glatte und spannungsfreie Verkupferung ermöglicht wird.
b. Es dürfen keine Glasfasern die Galvanisierung durchdringen.

Design Note
In order to avoid inquires from the PCB manufacturer regarding the meaning of the optical targets, the PCB designer should state on the master drawing the explanation "Optical targets for assembly" together with a representation of an optical target.

4.13 Electrical Testing of Finished Boards

The tests stated below are performed according to the customer's requirements or if necessitated by the manufacturing processes.

a. Electrical test for short and open circuits.
b. Automatic optical inspection (AOI)

Note
If the board is specified with masks, it may be necessary to perform the inspection prior to applying the masks.

5. Plated-Through Holes

5.1 General Requirements

a. Holes in the base material shall be drilled so that the requirements for soldering and unsoldering strength, Item 5.6, are fulfilled, i.e., the hole walls shall exhibit a certain unevenness. At the same time, the soldering conditions, Item 13.2, shall be fulfilled, i.e., the hole walls shall exhibit so small an unevenness that the copper plating becomes even and unstressed.

b. No glass fibres may protrude through the plating.

| Metallisierte Löcher | Plated-Through Holes |

c. Am Übergang von der Hülse zum Lötauge dürfen weder in der Kupfer- noch in der aufgebrachten Zinn-/Blei-Schicht Risse auftreten.

d. Die Löcher in fertigen Leiterplatten müssen frei von Schmutz, Staub, Bohrspänen, usw. sein und sich zuverlässig löten lassen. Vgl. Punkt 13.1.

5.2 Dicke des galvanischen Niederschlages

Siehe Punkt 2.2 und 2.3.

5.3 Durchmessertoleranz

Das fertige Loch hat den Nenndurchmesser d.

a. **Aufgeschmolzene/heißluftverzinnte Leiterplatten**

$d < 0{,}6$ mm: $\pm 0{,}10$ mm*
$0{,}6 \leq d \leq 2{,}0$ mm: $\pm 0{,}10$ mm
$2{,}0 < d \leq 6{,}3$ mm: $\pm 0{,}15$ mm
$d > 6{,}3$ mm: $\pm 0{,}20$ mm

* Gilt nur für die verkupferte Bohrung.

Es wird akzeptiert, daß sich Bohrungen mit d < 0,6 mm beim Aufschmelzen oder Heißluftverzinnen mit Zinn/Blei füllen.

b. **Engere Toleranzen**
Es können in Sonderfällen engere Toleranzen als die oben genannten auf der Bohrzeichnung angegeben werden. Diese Toleranzen können eventuell nur für die Verkupferung gelten.

5.4 Restringbreite

Als Restring wird die Fläche zwischen dem äußeren Rand des Pads und der Bohrung bezeichnet. Durch mehr oder weniger exzentrisches Bohren kann der Restring unterbrochen werden. Das hängt vom Pad-Außendurchmesser im Verhältnis zur Bauteileaufnahmebohrung ab (siehe Punkt 5.4 a bis 5.4 d). Die nominale Breite des Restringes bildet die Grundlage zur Beurteilung der fertigen Leiterplatte.

a. **Generelle Anforderungen**
Die Restringbreite t soll auf der fertigen Leiterplatte folgenden Anforderungen genügen:

Außenlagen: $t_1 \geq 0{,}05$ mm
Innenlagen: $t_2 \geq 0{,}01$ mm

c. The plated copper layer and the overlying tin/lead layer shall remain unbroken at the barrel-to-pad interface.

d. Holes in finished boards shall be free of dirt, dust, drill chips etc., and be fully solderable, cf. Item 13.1.

5.2 Plating Thickness

See Items 2.2 and 2.3.

5.3 Tolerance of Diameter

Finished hole with nominal diameter d.

a. **Reflowed/Hot-Air Levelled Boards**

$d < 0.6$ mm: ± 0.10 mm*
$0.6 \leq d \leq 2.0$ mm: ± 0.10 mm
$2.0 < d \leq 6.3$ mm: ± 0.15 mm
$d > 6.3$ mm: ± 0.20 mm

* Valid vor the copper plating only.

It will be accepted that holes < 0.6 mm become closed because of reflowing or hot-air levelling

b. **Tighter Tolerances**
Tighter tolerances than stated above may, in special cases, be stated on the master drawing, possibly valid for the copper plating only.

5.4 Annular Ring of the Solder Pad

The annular ring of an HMT solder pad is understood as the area lying between the outer rim of the solder pad and the solder hole. As a result of a more or less eccentric drilling, break-out of the annular ring can occur. This depends on the nominal outer diameter of the solder pad relative to the solder hole diameter. See Items 5.4 a to 5.4 d. The nominal width of the annular ring, forms the basis of the assessment of the finished board.

a. **General Requirement**
The width t of the annular ring shall fulfil the following requirements:

Outer layers: $t_1 \geq 0.05$ mm
Inner layers: $t_2 \geq 0.01$ mm

Plated-Through Holes

The width of the annular ring is measured as shown below.

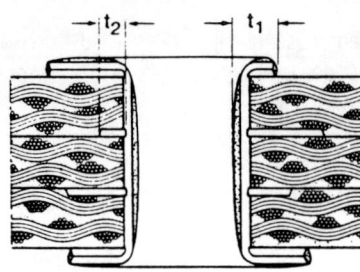

The outer diameter d_y of solder pads on outer layers as well as inner layers is determined by

$$d_y \geq d + 0.4 \text{ mm},$$

where d indicates the nominal diameter of the hole. The requirement is valid for board sizes up to 450 x 450 mm.

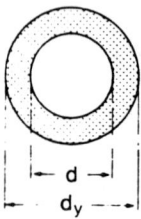

The requirement of unbroken annular rings cannot be maintained if

$$d_y < d + 0.4 \text{ mm}.$$

Design Note
For the sake of registration, it is recommended that d_y be increased by 0.1 mm, i.e.,

$$d_y \geq d + 0.5 \text{ mm}.$$

Apertures in Inner Layer Planes
It is implied that the diameter D of appertures in ground and voltage planes be determined by

$$D \geq d_y + 0.4 \text{ mm or}$$

$$D \geq d + 0.8 \text{ mm},$$

where d_y indicates the outer diameter of the pad and d the nominal diameter of the hole. This corresponds to a clearance (insulation distance) of 0.2 mm.

| Metallisierte Löcher | Plated-Through Holes |

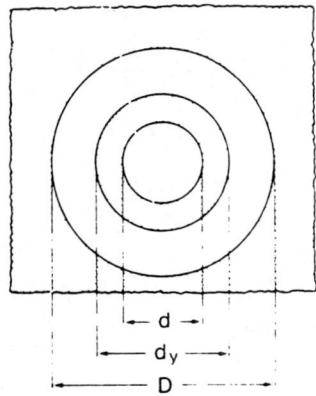

Mindestabstand zwischen Pad und Leiter auf den Innenlagen
Der Abstand A zwischen Padrand und Leiter soll den Forderungen entsprechen, die für die Aussparungen in Masse- und Spannungsebenen gelten, d.h.

$A \geq 0{,}2$ mm

Min. Pad-to-Conductor Distance on Inner Layers
It is implied that the distance A between the pad rim and the conductor observe the same requirements as stated for apertures in the ground and voltage planes, i.e.,

$A \geq 0.2$ mm

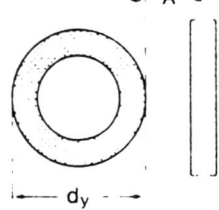

Designhinweis
Der genannte Mindestabstand ist auch dann einzuhalten, wenn d_y um 0,1 mm vergrößert wird. Vgl. obenstehenden Designhinweis.

b. Eine Unterbrechung des Restringes ist für 1 % der Lötaugen auf den Außenlagen zulässig, jedoch nicht dort, wo der Leiter angebunden ist. Der Mindestabstand a zwischen der Unterbrechung und dem Leiter soll 0,05 mm sein.

Design Note
The minimum distance stated shall also be fulfilled when d_y is increased by 0.1 mm, cf. the above design note.

b. Annular ring break-out is accepted at 1 % of the solder pads on the outer layers, however, not where the conductor is attached. The minimum distance a between the break-out and the conductor shall be 0.05 mm.

$a \geq 0.05$ mm
$n \leq 1\%$

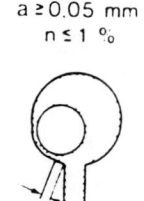

| Metallisierte Löcher | Plated-Through Holes |

Nicht zulässig
Unacceptable

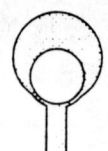

| c. Wenn der Restring unterbrochen ist, darf nicht mehr als 90° (d.h. ein Viertel) des Bohrlochumfanges (der Lochwand) außerhalb des Lötauges liegen. | c. In the case of break-out of the annular ring, no more than 90° of the outer periphery of the hole drilling (the hole wall) may fall outside the solder pad. |

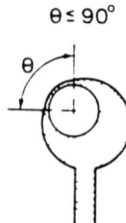

$\theta \leq 90°$

| d. Der tatsächliche Isolationsabstand R zwischen dem Bohrungsrand (der Lochwand) eines versetzten Loches und einem benachbarten Leiter darf nicht kleiner werden als 75 % des Nominalabstandes zwischen dem Lötauge und dem Leiter*, wobei die unter Punkt 4.2 bis 4.7 angegebenen Fehlermöglichkeiten zu berücksichtigen sind. | d. The actual insulation distance R between the outer periphery of the hole drilling (the hole wall) of a dislocated hole and an adjacent conductor shall not be reduced below 75 % of the nominal distance A between the solder pad and the conductor*, including the pattern faults stated in Items 4.2 to 4.7. |

Wenn der nominale Isolationsabstand bestimmt wird, ist in Betracht zu ziehen, daß sich, falls Punkt 5.4 c zutrifft, eine radiale Ausdehnung des Pads um bis zu 50 μm ergeben kann.

When determining the nominal insulation distance, it shall be taken into consideration that the radial extent of the break-out can amount to 50 μm when Item 5.4 c is fulfilled.

* Referenz: Leiterplattenspezifikation des Kunden, vgl. Punkt 4.1.

* Reference: Customer's documentation, cf Item 4.1.

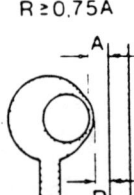

$R \geq 0{,}75 A$

| Metallisierte Löcher | Plated-Through Holes |

5.5 Fehlstellen in metallisierten Löchern

Kleine Fehlstellen sind unter folgenden Bedingungen zulässig:

a. Nicht mehr als 3 Fehlstellen in maximal 2 % der metallischen Bohrungen.

b. Die gesamte Fläche der Fehlstellen in einer Bohrung darf 5 % der gesamten Bohrungswandfläche nicht überschreiten.

5.5 Voids in Plated-Through Holes

Voids are accepted under the following conditions:

a. No more than 3 voids per solder hole in max. 2 % of the solder holes.

b. The total area of voids per solder hole shall not exceed 5 % of the area of the hole wall.

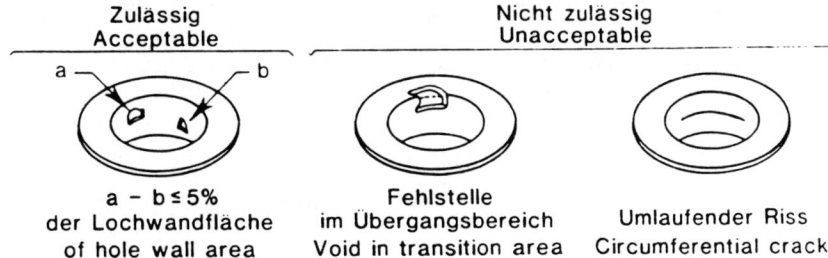

Zulässig / Acceptable: a − b ≤ 5% der Lochwandfläche / of hole wall area

Nicht zulässig / Unacceptable: Fehlstelle im Übergangsbereich / Void in transition area — Umlaufender Riss / Circumferential crack

c. Es dürfen keine Fehlstellen am Übergang Hülse-Pad (A) oder an der Kontaktierungsstelle innenliegender Pads (B) auftreten. Ferner dürfen keine Löcher an diametral gegenüberliegenden Stellen der Lochwand (C) vorkommen, da dies auf einen umlaufenden Riß schließen läßt.

c. No voids shall be found at the pad-to-barrel interface (A) or outside internal pads (B). Furthermore, voids shall not be found at the same level at diametrically opposite places of the hole wall (C) since this could indicate a circumferential crack.

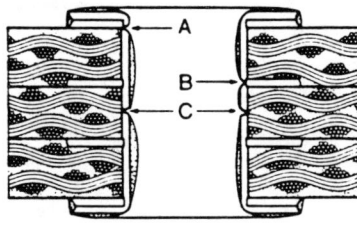

5.6 Löt- und Entlötbeständigkeit

Die Durchmetallisierung einer Bohrung muß eine so große Haftfestigkeit zum Basismaterial aufweisen, daß sie ein fünfmaliges Ein- und Auslöten eines Drahtes übersteht. Die Temperatur der Lötkolbenspitze soll zwischen 232 und 260 °C betragen. Die Wärme soll über den Draht und nicht direkt über das Pad übertragen werden. Siehe IPC-TM-650, Punkt 2.4.36.

5.6 Soldering/Unsoldering Strength

The through-plating shall exhibit sufficient adhesion to the base material to resist five times unsoldering and soldering a wire soldered into a plated-through hole. The temperature of the tip of the soldering iron shall lie within the interval 232 - 260 °C, and the heat shall be transferred via the wire, not directly to the solder pad. See IPC-TM-650, Item 2.4.36.

| Metallisierte Löcher | Plated-Through Holes |

5.7 Prüfabschnitt

Falls in der Spezifikation verlangt, sind Prüfabschnitte mitzuliefern. Hier kann man unter folgenden standardisierten Möglichkeiten wählen:

A: Es ist ein Prüfabschnitt pro Leiterplatte auszuliefern. Dieser ist dicht an der Leiterplatte angeordnet und so vorgefräst, daß er sich leicht abbrechen läßt.

B: Es ist ein Prüfabschnitt pro Nutzen auszuliefern. Hierbei sind sowohl der Prüfabschnitt als auch die einzelnen Leiterplatten so zu kennzeichnen, daß die Zusammengehörigkeit festgestellt werden kann.

C: Für jedes Fertigungslos sind ein Prüfabschnitt und Schliffbilder einschließlich eines Prüfberichtes mitzuliefern.

Die Kundenspezifikation (vgl. Punkt 4.1) kann den kundenspezifischen Prüfabschnitt enthalten, eventuell auch einen standardisierten Testabschnitt, der in Absprache mit dem Leiterplattenhersteller angebracht wird. Der Prüfabschnitt sollte gleichzeitig mit dem Leiterbild auf den Film geplottet werden. Es kann auch dem Leiterplattenhersteller überlassen werden, den Prüfabschnitt (eventuell seinen eigenen) auf dem Film anzulegen.

Anstelle von Prüfabschnitten, oder als Ergänzung, kann gefordert werden, daß eine Ausschußplatte zusammen mit dem Los geliefert wird, so daß der Kunde seine eigenen Schliffuntersuchungen vornehmen kann. Fehler, die das Schliffbild beeinträchtigen können, sind kein Rückweisungsgrund für die Leiterplatte. Um Verwechslungen zu vermeiden, sollte eine Ecke der Ausschußplatte abgeschnitten werden.

Die Überprüfung der im folgenden beschriebenen Punkte 5.8 und 5.20 wird anhand von Schliffbildern durchgeführt. Die unter 5.8 bis 5.17 beschriebenen Schliffbilder gelten für Bohrungen ohne vorausgehende Beanspruchung, während die Schliffe von Punkt 5.18 bis 5.20 nach einem Lötprozeß bei 288 ± 6 °C bei 10 s Lötdauer gemacht werden.

Hinweis
Ein gutes Testergebnis für einen Prüfabschnitt sollte nicht als schlüssiger Beweis für Fehlerfreiheit einer Leiterplatte genommen werden. Andererseits sollte ein schlechtes Testergebnis als Anzeichen für eine unzureichende Leiterplattenqualität gelten.

5.8 Epoxid-Smear (Verschmierung)

Es darf keine Epoxidverschmierung e oder andere Trennung zwischen der Kupferhülse und den innenliegenden Pads auftreten.

5.7 Test Coupon

When required on the master drawing, test coupons shall be delivered. The following choices are standardized:

A: One test coupon per board shall be delivered. The test coupon shall be tabattached to the board, i.e. contoured in such a way that it is easy to detach.

B: One test coupon per production panel shall be delivered. The test coupon and the individual boards of the production panel shall be marked in a way to ensure traceability.

C: One test coupon and a microsection report, including photos, are to be delivered with each batch manufactured.

The customer's documentation of the board, cf. Item 4.1 may include the customer's own test coupon, possibly a standard test coupon located in agreement with the PCB manufacturer. The test coupon should be plotted on the films simultaneously with the pattern. It may also be left to the PCB manufacturer to locate the test coupon, or, the PCB manufacturer's own test coupon, on the films.

Instead of test coupons, or possibly as a supplement, it may be specified that a scrap board be co-delivered so that the customer can prepare microsections of his own. The board shall not be rejected because of defects compromising the microsections. To avoid mix-up with good boards, one corner of the rejected board should be cut off.

Verification of Items 5.8 through 5.20 below is performed on the basis of microsections. The microsections in Items 5.8 through 5.17 are prepared with no pre-treatment whereas the microsections in Items 5.18 through 5.20 are prepared after a preceding soldering of the holes at 288 ± 6 °C for 10 sec.

Note
A good test result of a test coupon should not be taken as conclusive evidence of a satisfactory quality of the board. Conversely, a poor test result should be considered a sign of an unsatisfactory quality of the board.

5.8 Epoxy Smear

No epoxy smear e, or any other separation, shall be found between the plated hole wall and inner pads.

| Metallisierte Löcher | Plated-Through Holes |

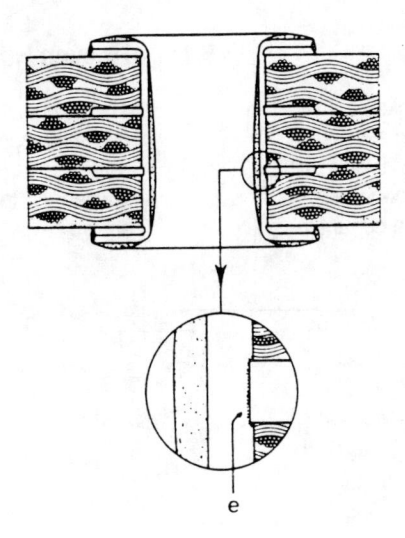

5.9 Rückätzung des Basismaterials

Eine Rückätzung ist nicht erforderlich. Falls der Leiterplattenhersteller eine Rückätzung durchführen will, sollte deren Tiefe t < 75 μm sein.

5.9 Etchback of Base Material

Etchback is not required. Should the PCB manufacturer wish to perform an etchback, the depth t of the etchback shall be less than 75 μm.

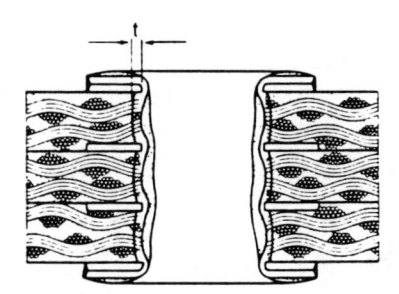

5.10 Rückätzen des Kupfers

Falls das Kupfer der außen- oder innenliegenden Pads in der Bohrung bei einem Ätz- oder Reinigungsprozeß angegriffen wird, dann darf es sich nur um t ≤ 10 μm zurückziehen.

Das gilt dann, wenn nicht, wie unter Punkt 5.9 beschrieben, rückgeätzt wird.

5.10 Etchback of Copper

A possible withdrawal t of inner and outer pads relative to the glass epoxy wall, e.g., because of an etching or cleaning process, shall not exceed 10 μm.

It is implied that no etchback be required as per Item 5.9.

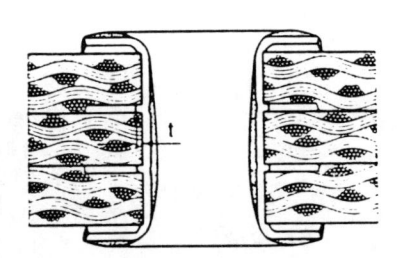

Metallisierte Löcher	Plated-Through Holes
5.11 Unebene Lochwand	**5.11 Uneven Hole Wall**
Die Lochwand darf keine größere Welligkeit aufweisen als durch eine maximale Rückätzung des Basismaterials von 75 μm hervorgerufen wird. Das gilt sowohl für eine ebene als auch unebene Rückätzung. Vgl. die linke und die rechte Seite der untenstehenden Skizze.	The hole wall shall not exhibit a greater waviness d than corresponding to a maximum etchback of 75 μm. This applies whether the etchback is even or uneven, cf. the left and right sides of the illustration shown below.

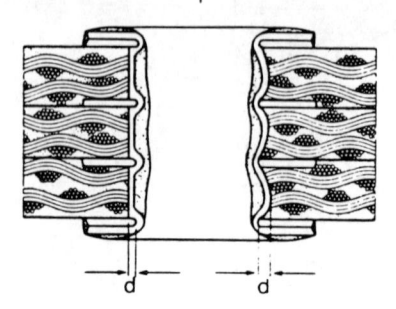

5.12 Porosität in der gebohrten Lochwand	**5.12 Porosity in Drilled Hole Wall**
Poren und Kanäle in der gebohrten Lochwand, die sich z.B. bei der Lochwandreinigung, dem Rückätzen und ähnlichen Prozessen bilden können, dürfen, gemessen von der Bohrungswand aus, nur maximal 0,1 mm tief sein.	Porosity or radial channels in the base material around the drilled hole wall, e.g., because of cleaning of epoxy smear or etchback and the like, shall not extend into the base material by more than 0.1 mm from the drilled hole wall.

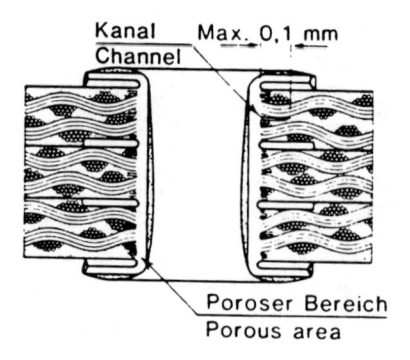

5.13 Plating Pockets ("Taschen" in der Galvanikschicht)	**5.13 Plating Pockets**
Beim Galvanisieren entstehende "Taschen" in der Hülse sind unter folgenden Voraussetzungen zulässig:	On the condition that all the requirements stated below are met, plating pockets* are acceptable:
a. Die Lochwand muß ansonsten glatt und zusammenhängend sein.	a. The hole wall shall exhibit an otherwise smooth and coherent structure.
b. Die Umgebung einer Tasche darf keine Risse aufweisen.	b. The region around a plating pocket shall not show any sign of cracking.
c. Es dürfen sich keine Taschen in Höhe der Innenlagenanschlüsse befinden.	c. No plating pockets may be found on a level with the interface between the plated hole wall and the pads of the inner layers.

Metallisierte Löcher	Plated-Through Holes
d. Die restliche Kupferdicke zwischen Taschengrund und gebohrter Lochwand muß der in Punkt 2.2 a genannten Anforderung genügen. Wenn nicht, ist die Tasche als Fehlstelle zu betrachten, die nach Punkt 5.5 zu beurteilen ist.	d. The remaining copper layer as measured from the bottom of a plating pocket to the drilled hole wall shall fulfil the requirement in Item 2.2 a regarding thickness. If not, the plating pocket is regarded as a void which is assessed in accordance with Item 5.5.

• (A) Offene Tasche
  (B) Geschlossene Tasche
	• (A) Open plating pocket
	  (B) Closed plating pocket

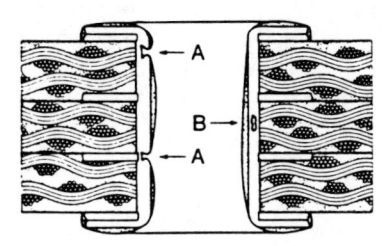

5.14 Knospen / Nodules

Knospen sind akzeptabel, wenn folgende Bedingungen erfüllt sind:	On the condition that all the requirements stated below are met, nodules are acceptable:
a. Die Lochwand muß ansonsten glatt und zusammenhängend sein (A).	a. The hole wall shall exhibit an otherwise smooth and coherent structure (A).
b. Der Lochdurchmesser entspricht den unter Punkt 5.3 genannten Anforderungen.	b. The diameter of the hole meets the requirements stated in Item 5.3.
c. Die Umgebung einer Knospe darf keine Anzeichen eines Risses aufweisen (B). Siehe auch Punkt 5.18.	c. The region around a nodule shall not show any sign of cracking (B). See also Item 5.18.

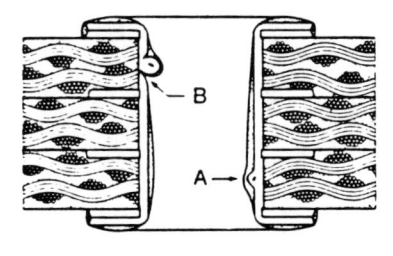

5.15 Nagelkopfbildung / Nailheading

Falls eine Nagelkopfbildung auftritt, darf die Kupferdicke nur um maximal 50 % ansteigen.	Nailheading with a thickening greater than 50 % of the thickness a of the copper shall not be found.

| Metallisierte Löcher | Plated-Through Holes |

| Die Höhe des Kupfergrates darf bis zur halben Dicke der Kupferfolie, höchstens jedoch 25 µm betragen. Die Grathöhe g wird vom Fuß bis zur Spitze gemessen (A), auch wenn der Grat in die Bohrung zurückgedrückt wurde (B). | Copper burrs at the edge of the drilled hole shall not exhibit a burr heigth greater than 50 % of the thickness of the copper foil, maximum 25 µm. The burr height is the dimension from its foot to its top (A), even when the burr is pressed back into the hole (B). |

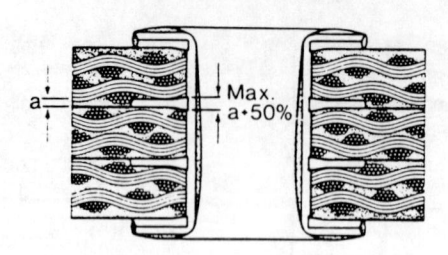

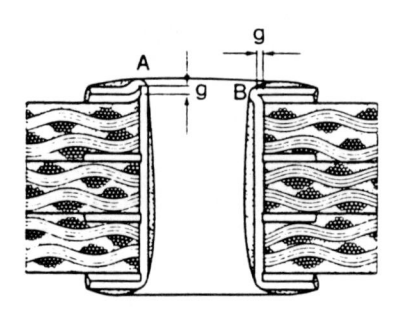

5.17 Ablösung der Hülse vom Laminat | 5.17 Laminate-Hole Wall Separation

| In einer unbeanspruchten Leiterplatte, d.h. im Anlieferungszustand, darf sich eine Ablösung des chemischen Kupfers vom Laminat maximal über 40 % der Leiterplattendicke t erstrecken. Das gilt für beide Seiten eines Schliffes. | In an unstressed board, i.e. as received, a separation between the laminate and the electroless copper of the hole wall shall not exceed 40 % of the board's thickness t. This applies to each side of the microsection. |

Die Ablösung kann auftreten durch

a. Vertiefungen in der Lochwand (A)
b. Ausbeulungen in der Hülse (B)

The separation can occur as:

a. Resin recession around the hole wall (A).
b. Hole wall pull-away (B).

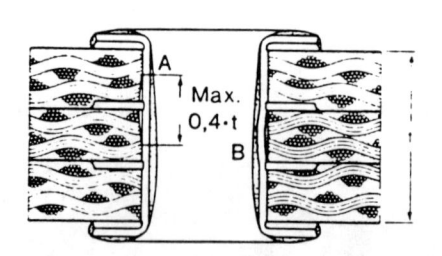

| Metallisierte Löcher | Plated-Through Holes |

5.18 Risse in der Metallisierung

Risse sind nicht zulässig. Risse können in folgender Form auftreten:

a. Als Risse an der Lochkante (A) und (B), eventuell über den ganzen Umfang.
b. Umlaufende Risse in der Hülse (C).
c. Risse in der Nähe von Taschen (E) und Knospen (D).
d. Risse in einem innenliegenden Pad oder einer Innen-Kupferlage (F).

Bemerkung
Vor der Kontrolle auf Risse sind die Leiterplatten unter den in Punkt 5.7 genannten Bedingungen zu löten.

5.18 Cracks in the Plating

Cracks are not acceptable. Examples of cracks are:

a. Corner cracks (A) and (B), possibly circumferential.
b. Circumferential barrel crack (C).
c. Crack around nodule (D) and plating pocket (E).
d. Crack in inner pad or inner copper layer (F).

Note
A preceding soldering in accordance with the soldering conditions stated in Item 5.7 is implied.

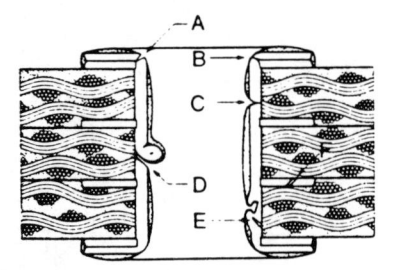

5.19 Verbindung zwischen stromlos und galvanisch abgeschiedenem Kupfer

Hülse
Zwischen der stromlos abgeschiedenen (a) und der galvanischen (b) Kupferschicht muß eine vollständige Verbindung bestehen.

Bemerkung
Vor den Schliffuntersuchungen ist gemäß Punkt 5.7 zu löten.

5.19 Plating Contact

Hole Barrel
Full plating contact between the electroless copper layer (a) and the electrolytic copper layer (b) of the hole barrel required.

Note
A preceding soldering in accordance with the soldering conditions stated in Item 5.7 is implied.

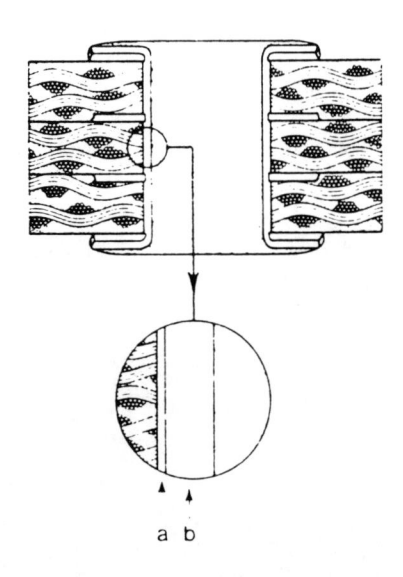

| Metallisierte Löcher | Plated-Through Holes |

Innenlagen

Zwischen dem stromlos abgeschiedenen Kupfer (b) in der Bohrung und der Innenkante der inneren Pads (c) muß über mindestens 90 % des Bohrungsumfanges ein vollständiger Kontakt (A) bestehen. Dies kann mit Hilfe eines Horizontalschliffes überprüft werden. Das galvanisch abgeschiedene Kupfer ist mit (d) bezeichnet. Ein partieller Kontakt, wie bei (B) und (C) dargestellt, ist nicht zulässig.

Bemerkung

Vor der Herstellung des Schliffes sind die Leiterplatten entsprechend Punkt 5.7 zu löten.

Inner Layers

Full plating contact (A) over minimum 90 % of the hole circumference, as seen on a horizontal microsection, is required between the electroless copper layer (b) of the hole plating and the edge of the inner pads (c). (d) indicates the electroplated copper. Partial plating contact (B) and (C) is not acceptable.

Note

A preceding soldering in accordance with the soldering conditions stated in Item 5.7 is implied.

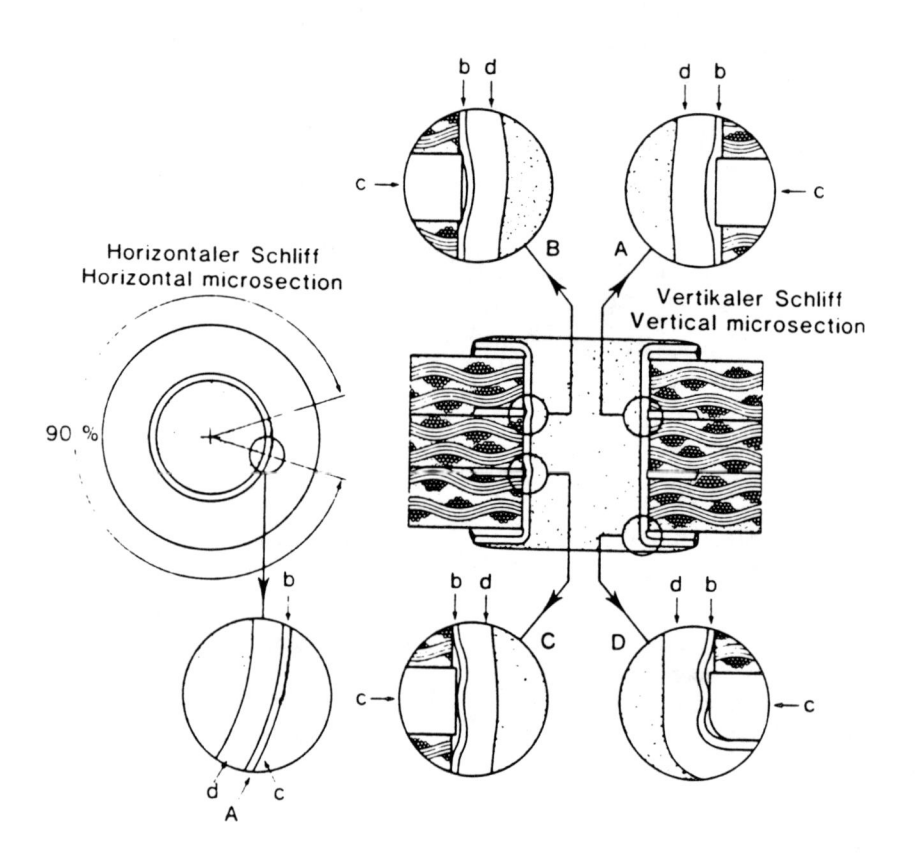

Außenlagen

Ein vollständiger Kontakt zwischen dem stromlos abgeschiedenen Kupfer der Lochwandmetallisierung (b) und der Innenkante der Pads auf den Außenlagen (c) ist nicht erforderlich. Ein partieller Kontakt ist zulässig (D).

Bemerkung

Vor der Herstellung des Schliffes sind die Leiterplatten entsprechend Punkt 5.7 zu löten.

Outer Layers

Full plating contact between the electroless copper layer of the hole plating (b) and the entire edge of the outer pads (c) is not required. A partial plating contact is acceptable (D).

Note

A preceding soldering in accordance with the soldering conditions stated in Item 5.7 is implied.

| Nicht metallisierte Löcher | Nonplated-Through Holes |

5.20 Delaminierung

Die fertige Leiterplatte darf keine als Folge des Lötens entstehende oder wachsende Delaminierung aufweisen. Als Delaminierung wird eine unerwünschte Trennung zwischen den einzelnen Schichten verstanden, aus denen eine Leiterplatte besteht: Prepreg, starres Laminat und Kupfer.

(A) zeigt eine Trennung zwischen den einzelnen Schichten des Basismaterials, (B) einen Hohlraum in diesen. Bei (C) und (D) ist eine Trennung zwischen dem Laminat und den Innen- sowie Außenpads beziehungsweise der Masse oder Versorgungsebene erkennbar (C).

Kleine, zufällig auftretende Hohlräume mit maximal 100 µm Größe sollten nicht als Fehler betrachtet werden, falls sie nicht beim Löten entstehen.

Bemerkung
Vor den Schliffuntersuchungen ist gemäß Punkt 5.7 zu löten.

5.20 Delamination

The finished board shall not exhibit delamination which develops (grows) because of soldering. By delamination is understood an unwanted separation between the individual layers making up the board: prepregs, rigid laminate and copper.

(A) shows a separation between the individual layers, and (B) shows a cavity in these. (C) and (D) show a separation between the laminate and inner and outer pads, respectively, alternatively ground and voltage planes (C).

Small cavities occurring at random, with a maximum size of 100 µm should not be regarded as a flaw if they do not develop during soldering.

Note
A preceding soldering in accordance with the soldering conditions stated in Item 5.7 is implied.

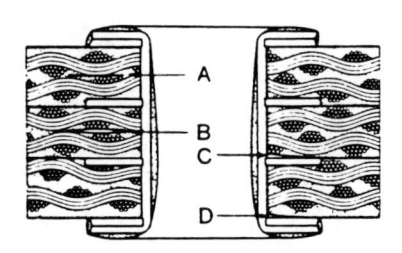

6. Nicht metallisierte Löcher

Nicht metallisierte Löcher können als Bauteilbohrungen oder als Montagebohrungen vorkommen. Im allgemeinen haben Montagebohrungen keine Pads. Wenn es dem Leiterplattenhersteller freigestellt wird, aus Kostengründen Montagebohrungen zu metallisieren, so sollte das in der Spezifikation angegeben werden. Pads dürfen jedoch nur nach Rücksprache mit dem Leiterplattendesigner an Montagebohrungen angebracht werden.

6. Nonplated-Through Holes

Nonplated-through holes can occur, partly as solder holes, partly as mounting holes. In general, mounting holes will have no pads. Where the PCB manufacturer for reasons of economy is allowed a free choice to execute mounting holes as plated-through holes, this should be stated on the master drawing. However, pads can only be added at mounting holes after consulting the PCB designer.

6.1 Allgemeine Qualitätsanforderungen

a. Lötaugen dürfen sich nicht vom Basismaterial abheben.

6.1 General Requirements

a. Solder pads shall not lift from the base laminate.

| Nicht metallisierte Löcher | Nonplated-Through Holes |

b. Eventuell auftretender Grat muß an der Ober- wie an der Unterseite einer Bohrung kleiner als 50 µm sein.

c. Eine volle Abdeckung des Lochrandes mit Zinn/Blei ist nicht erforderlich.

6.2 Durchmessertoleranzen

Fertige Bohrung mit Nenndurchmesser d.

$0{,}6 \leq d \leq 6{,}3$ mm: ± 0,10 mm
$d > 6{,}3$ mm: ± 0,20 mm

6.3 Restring

Aus löttechnischen Gründen darf der Restring von nicht metallisierten Bohrungen nicht unterbrochen sein. Das setzt eine Mindestringbreite voraus. Die aus den Fertigungsunterlagen hervorgehende nominale Ringbreite ist die Grundlage zur Beurteilung der fertigen Leiterplatte.

Die untenstehenden Bedingungen gelten nur, wenn nicht metallisierte Bohrungen mit Lötaugen versehen sind und zum Einlöten von Bauteilen benutzt werden.

Die Restringbreite t darf auf der fertigen Leiterplatte an keiner Stelle kleiner als 0,15 mm sein.

---

b. The height of possible burrs at the top and bottom sides of the holes shall be less than 50 µm.

c. Full coverage of the hole edge with tin/lead is not required.

6.2 Tolerance of Diameter

Finished hole with nominal diameter d.

$0{.}6 \leq d \leq 6{.}3$ mm: ± 0.10 mm
$d > 6{.}3$ mm: ± 0.20 mm

6.3 Annular Ring

For soldering reasons, nonplated-through holes shall not exhibit annular ring break-out. This implies that the annular rings be designed with a certain minimum width. The nominal width of the annular ring which appears in the documentation, forms the basis of the assessment of the finished board.

The requirements below are valid only in such cases where nonplated-through holes are provided with solder pads and used as solder holes.

The width t of the annular ring shall nowhere on the finished board be less than 0.15 mm.

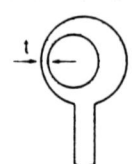

$t \geq 0.15$ mm

Dabei ist vorausgesetzt, daß der Außendurchmesser des Lötauges mit

$d_y \geq d + 0{,}8$ mm

definiert ist, wobei d der Nenndurchmesser der Bohrung ist. Die Anforderung gilt für Leiterplattengrößen bis 450 x 450 mm.

It is implied that the outer diameter d_y of the pad is determined by

$d_y \geq d + 0.8$ mm

where d indicates the nominal diameter of the hole. The requirement is valid for board sizes up to 450 x 450 mm.

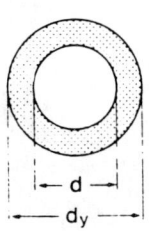

| Vergoldete Kontakte | Gold Plated Contacts |

7. Vergoldete Kontakte

(Gilt auch für Schalterkontakte, Tastaturen und ähnliches.)

7.1 Allgemeine Qualitätsanforderungen

a. Die Oberfläche muß eben und kratzerfrei sein und von Kontakt zu Kontakt ein einheitliches Aussehen aufweisen.

b. Alle Kanten bis auf das bearbeitete Ende müssen mit Gold bedeckt sein.

Designhinweis
Die elektrische Anbindung an den Galvaniksteg sollte von den Kontakten aus unsymmetrisch angelegt werden.

c. Wenn nach der Leiterplattenspezifikation selektiv vergoldet wird, bleiben die Seitenkanten frei von Gold.

d. Die aufgalvanisierten Schichten (Au oder Ni + Au) dürfen sich nicht abschälen. Siehe Abschnitt 7.5.

e. Angrenzende Flächen mit Zinn/Blei-Überzug ("A" siehe Skizze) dürfen nicht beschädigt werden.

7. Gold Plated Contacts

(Also valid for switch contacts, keyboard contacts and the like.)

7.1 General Requirements

a. The surface must be smooth, with no scratches and with an uniform appearance from contact to contact.

b. All edges but the machined end of the contacts shall be covered with gold.

Design Note
The electrical connection to the plating bar should be placed unsymmetrically with respect to the contacts.

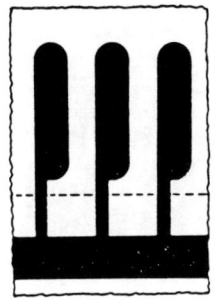

c. If the master drawing states that gold may be used as an etch resistant, noncoverage of the side edges with gold is accepted.

d. No flaking of the plated layer(s) (Au or Ni + Au) shall be found. See Item 7.5.

e. The tin/lead plating of possible text (A) in adjacent areas shall not be damaged.

Galvaniksteg
Plating bar

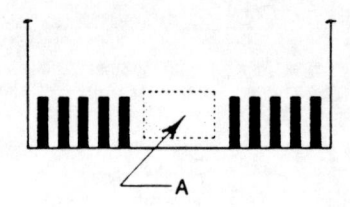

Vergoldete Kontakte	Gold Plated Contacts
f. Der Übergang vom vergoldeten Kontakt zum Leiterzug muß so ausgeführt werden, daß keine Korrosion auftreten kann.	f. The interface between the gold plated contacts and the attached conductors shall be made in such a way that no corrosion can occur.
g. Die unter den Punkten 7.1 a, b und d sowie 7.3 und 7.4 genannten Anforderungen gelten nicht für unbenutzte Kontakte.	g. The requirements stated in Items 7.1 a, b and d, and also 7.3 and 7.4 are not valid for unutilized contacts.
h. Der Abstand d zwischen dem oberen Rand eines Kontaktfingers und dem Rand des nächstgelegenen Lötauges muß mindestens 3 mm betragen.	h. It is implied that the board be designed with a distance d of min. 3 mm between the top of the contacts and the rim of the nearest solder pads/via holes.

d ≥ 3 mm

7.2 Galvanisierung

Schichtdicke und Materialien sind unter Punkt 2.4 beschrieben.

7.2 Plating

Plating thickness and materials are stated in Item 2.4.

7.3 Nadellöcher (Pinholes)

Nadellöcher sind unter folgenden Voraussetzungen zulässig:

a. Der Nadellochdurchmesser muß kleiner als 0,10 mm sein.

b. Es darf nicht mehr als 1 Nadelloch pro Kontakt auftreten, und zwar nur außerhalb der effektiven Kontaktfläche (A). Die betreffende Fläche ist ein um den ganzen Kontakt führender Randstreifen, dessen Breite 10 % der Kontaktbreite beträgt.

7.3 Pinholes

Pinholes are accepted under the following conditions:

a. The diameter of the pinhole is less than 0.10 mm.

b. No more than one pinhole per contact shall be found, and it shall be located outside the effective contact area (A). This lies within a band of 10 % of the contact width all way along the contact edge.

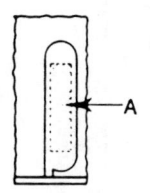

7.4 Porosität

Die galvanisch abgeschiedene Schicht muß so dick sein, daß sie einen Porositätstest nach IPC-TM-650, Punkt 2.3.24, besteht. Die Poren müssen sich außerhalb der effektiven Kontaktfläche befinden. Vgl. Punkt 7.3 b.

7.5 Haftfestigkeit der galvanischen Beschichtung

Die Haftfestigkeit muß ausreichen, um den Klebebandtest nach IPC-TM-650, Punkt 2.4.1 zu bestehen. Als alternatives Klebeband kommt Typ 691 von 3M Company in Frage.

7.6 Ausrichtung von Ober- zu Unterseite

Bei doppelseitigen Kontakten ist ein Versatz von 0,2 mm zwischen Ober- und Unterseite akzeptabel.

7.4 Porosity

The plating shall be sufficiently pore-free to fulfil a porosity test according to IPC-TM-650, Item 2.3.24. Pores shall be located outside the effective contact area. Cf. Item 7.3 b.

7.5 Plating Adhesion

The adhesion shall be sufficient to fulfil the tape test according to IPC-TM-650, Item 2.4.1. An alternative tape is type 691 from the 3M Company.

7.6 Side-to-Side Register

In the case of double-sided contacts, a misregistration of 0.2 mm between the contacts of the two sides of the board can be accepted.

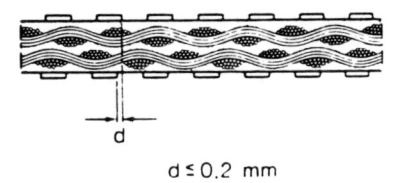

$d \leq 0.2$ mm

8. Siebgedruckte Masken und Komponentenbezeichnungen

8.1 Definition

Für SMD-Leiterplatten oder Leiterplatten mit Mischbestückung sind die Begriffe Lötseite, Bestückungsseite, Lötstoppmaske und Isoliermaske nicht länger eindeutig. Daher wird in den folgenden Ausführungen nicht mehr zwischen den einzelnen Begriffen unterschieden, sondern nur noch die Bezeichnung "Maske" verwendet. Die Anordnung des Masken- und Komponentendruckes geht aus den Fertigungsunterlagen hervor.

8. Screen-Printed Masks/Components Notations

8.1 Definition

In the case of surface mounted boards or mixed technology boards, the concepts solder side, component side, solder mask and insulation mask are no longer unambiguous. Therefore, the specifications below do not distinguish between the said concepts, the only designation being "mask". The locations of masks and component notations appear from the master drawing.

Für Leiterplatten für die Lochmontagetechnik gelten nach wie vor die genannten Begriffe, auch wenn sie nicht direkt in den Spezifikationen angeführt sind.

In the case of hole mounted boards the above concepts are still valid, although they are not directly stated in the specifications.

8.2 Einsatzbereich

Der Siebdruck wird zum
- ein- oder zweiseitigen Auftragen von Masken und
- Bauteilebezeichnungen (Komponentendruck)

benutzt.

8.2 Extent

Screen-printing is used for applying:
- Masks on one or both sides of the board
- Component notations

8.3 Allgemeine Anforderungen

a. Siebgedruckte Masken und Bauteilebezeichnungen müssen gut auf der Leiterplatte haften und dürfen nicht durch die beim Maschinenlöten auftretende Wärme beschädigt werden (siehe Punkt 13.2).

   Maskendruck auf Kupfer
   Die Maske darf sich nicht kräuseln, ablösen oder Haarrisse bilden. Die Haftfestigkeit ist nach IPC-TM-650, Punkt 2.4.28.1 zu prüfen. Alternativ kann das Klebeband Typ 691 von 3M Company verwendet werden.

   Maskendruck auf Zinn/Blei
   Ein leichtes Kräuseln der Maske nach dem Maschinenlöten kann akzeptiert werden. Die Maske muß sich der Struktur des erstarrten Zinn/Bleis anpassen. Wenn die abgedeckten Leiter schmaler als 1,27 mm sind, darf kein Ablösen auftreten. Dies wird nach IPC-TM-650, Punkt 2.4.28.1 geprüft. Alternativ kann Klebeband Typ 691 von 3M Company verwendet werden.

b. Eine gewisse Druckunschärfe aufgrund des Siebgewebes und ein eventuelles Ausbluten von pigmentfreiem Siebdrucklack vor dem Aushärten sind unzulässig. Ausbluten ist besonders dann erkennbar, wenn die Maske auf blankes Kupfer gedruckt wurde.

c. Es darf kein unerwünschter Siebdrucklack auf den Pads oder in den Durchmetallisierungen vorkommen.

d. Der Lack muß resistent gegen gebräuchliche Lösungsmittel sein: gegen chlorierte und fluorierte Kohlenwasserstoffe, Isopropanol und ähnliches, sowie gegen allgemein übliche Flußmittel.

e. Unerwünschte Metallpartikel wie z.B Bohrspäne dürfen weder in der Maske eingeschlossen sein noch auf deren Oberfläche haften.

8.3 General Requirements

a. Screen-printed masks and component notations shall adhere well to the board and shall not be damaged by heat when the board is mass soldered as specified in Item 13.2.

   Bare Copper under the Mask
   No wrinkling, crazing or flaking of the mask shall be found. The adhesion shall be tested according to IPC-TM-650, Item 2.4.28.1. An alternative tape is type 691 from the 3M Company.

   Tin/Lead under the Mask
   A slight wrinkling of the mask after mass shall follow the structure of the solidified tin/lead. No flaking shall occur where the underlying conductors are narrower than 1.27 mm. This shall be tested according to IPC-TM-650, Item 2.4.28.1. An alternative tape is type 691 from the 3M Company.

b. Some blur in the print due to the screen fabric and a possible bleeding of pigmentfree ink before curing will be accepted. Bleeding is particularly visible on boards with bare-copper under the mask.

c. Unintentional ink (ink film) shall not be found on the pads or in the plated-through holes.

d. The ink shall be resistant to solvents generally used: chlorinated and fluorinated hydrocarbon, isopropanol and the like, and also commonly used fluxes.

e. Unwanted metal particles, e.g., drill chips, shall not be found as inclusions in the mask or adhering to the surface of the mask.

Siebgedruckte Masken und Komponentenbezeichnungen		Screen-Printed Masks/Components Notations	

f. Durch ionische oder andere Verunreinigungen, die unter der Maske vorhanden sein könnten, darf keine wesentliche Verschlechterung der elektrischen Eigenschaften der Leiterplatte eintreten. Der Isolationswiderstand zwischen zwei beliebigen Knotenpunkten muß bei einer Prüfspannung von 100 VDC mindestens 500 MΩ betragen.

g. Die Maske muß vollständig ausgehärtet sein. Eine unvollständige Aushärtung kann beim Maschinenlöten zur Netzbildung (Webbing) führen, was nicht akzeptiert werden kann. Webbing kann jedoch auch durch unkorrekte Lötparameter wie z.B. fehlerhaftes Fluxen verursacht werden. Daher muß der Verarbeiter die Ursachen ergründen, z.B. durch Löten von Leiterplatten aus anderen Lieferchargen.

h. Es darf keine Delaminierung (Lufteinschlüsse unter der Maske) auftreten.

Designhinweis
Um ein Ablösen der Maske bei der Bearbeitung zu vermeiden, soll der Abstand zwischen Maskenrand und Leiterplattenrand normalerweise mindestens 0,5 mm betragen.

i. In der Maske dürfen nicht mehr als 2 Pinholes (Nadellöcher) pro mm^2 vorkommen. Pinholes sind Löcher mit einem größten Maß von 0,1 mm.

8.4 Materialien

Wenn kein bestimmter Lacktyp vorgeschrieben ist, können die folgenden Typen verwendet werden:

> thermisch härtende Lacke
>
> UV-härtende Lacke

8.5 Ausführung

a. Die durchschnittliche Schichtdicke einer Lötstopp- oder Isoliermaske, gemessen über eine ununterbrochene Fläche von 50 x 50 mm (Kupfer- oder Basismaterial) darf nicht geringer sein als

> 15 μm für thermisch härtende Lacke
>
> 10 μm für UV-härtende Lacke

Größere Schichtdicken lassen sich durch zweimaliges Drucken erzielen. Wenn notwendig, ist dies in den Fertigungsunterlagen anzugeben.

---

f. Neither ion-contamination nor other contamination shall be found under the mask to such a degree that an essential reduction of the board's electrical characteristics occurs. The insulation resistance between two arbitrary nodal points shall be ≥ 500 MΩ when applying a test voltage of 100 V dc.

g. The mask shall be fully cured. Occurrence of webbing after mass soldering is not acceptable if caused by incomplete curing. Webbing, however, can also be caused by inexpedient soldering parameters, e.g., incorrect fluxing, in which case the user should clarify the reason, e.g., by soldering boards from other deliveries.

h. No delamination, i.e., air pockets under the mask, shall be found.

Design Note
In order to avoid delamination during machining, the distance between the board edges and applied mask edges shall normally not be less than 0.5 mm. This applies also to internal cut-outs.

i. Pinholes in a mask shall not occur with a density higher than 2 pinholes per mm^2. Pinholes are apertures with a maximum dimension of 0.1 mm.

8.4 Materials

Unless a particular type of ink is specified, the following types can be used:

> thermally curable inks
>
> UV-curable inks

8.5 Execution

a. The average thickness of a solder or insulation mask, measured across an unbroken area (copper or base material) of 50 x 50 mm, shall not be less than:

> 15 μm for thermally curable inks
>
> 10 μm for UV-curable inks

Higher thicknesses can be obtained by double-printing. When required, this should be stated on the master drawing.

b. Die Komponentenbezeichnung kann entsprechend den Fertigungsunterlagen auf beide Leiterplattenseiten gedruckt werden, und zwar normalerweise auf die vorher aufgetragene Maske.

b. The component notation can be screen-printed on one or both sides of the board according to the master drawing, usually on top of the masks.

8.6 Überdrucken von Lötflächen (Pads)

Beim Maskendruck ist ein Überdecken von Lötflächen nicht zulässig, wenn die Aussparungen in den Masken folgenden Anforderungen genügen:

$d \geq d_y + 0{,}5$ mm   für Lötaugen bei der Lochmontagetechnik
$L \geq l + 0{,}5$ mm   bei SMT-Lötflächen
$B \geq b + 0{,}5$ mm   bei SMT-Lötflächen

Das bedeutet, daß ein Mindestabstand von 0,25 mm zwischen Padkante und Aussparungsrand einzuhalten ist. Die Bezeichnungen gehen aus der nachfolgenden Skizze hervor.

8.6 Overprinting of Solder Pads

At masks, overprinting of solder pads with ink will not be accepted when the aperture in the mask fulfils the requirement:

$D \geq d_y + 0.5$ mm   for HMT solder pads
$L \geq l + 0.5$ mm   for SMT solder pads
$B \geq b + 0.5$ mm   for SMT solder pads

This corresponds to a distance a (clearance) of min. 0.25 mm between the pad and the aperture edge. The designations appear from the illustration below.

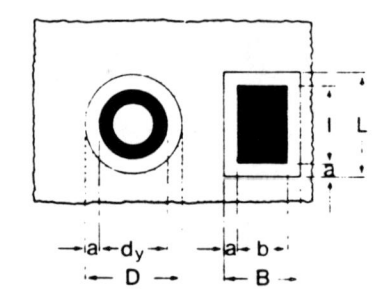

Designhinweis

Eine siebgedruckte Bauteilebezeichnung (z.B. C8) soll den gleichen Abstand a vom Pad einhalten, der auch für die Maskenaussparung gilt.

Um zu gewährleisten, daß der Maskendruck nicht über den Aussparungsrand einer schlecht ausgerichteten Maske hinausgeht, sollte der Mindestabstand zwischen Pad und Komponentendruck gleich dem Lötmaskenabstand + 0,4 mm sein.

Design Note

A screen-printed component notation (e.g., C8) shall keep the same distance a (clearance) from the solder pad as stated above for the aperture in the screen-printed mask.

To ensure that the component notation cannot be applied beyond the aperture edge in a misregistered mask, the min. distance a between the solder pad and the component notation should be the solder mask clearance + 0.4 mm.

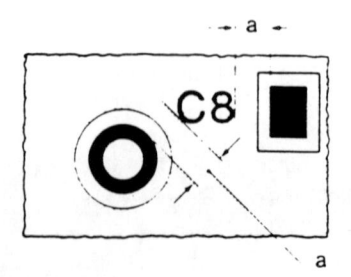

| Siebgedruckte Masken und Komponentenbezeichnungen | Screen-Printed Masks/Components Notations |

Dies gilt für Leiterplatten bis zu 450 x 450 mm. Bei größeren Leiterplatten und zweimaligem Maskendruck kann eine gewisse Überlappung auftreten.

It is implied that the board size is limited to 450 x 450 mm. In case of large boards, the requirement may imply printing twice with a certain overlapping.

8.7 Bedeckung von Leitern und Ausfüllung von Zwischenräumen

a. Ein an einem Pad vorbeiführender Leiter muß vollständig von der Maske bedeckt sein, vorausgesetzt, daß der Nominalabstand zwischen Leiter und Maskenausschnitt größer als 0,25 mm ist. Bei einem Zwischenraum zwischen Pad und Maske von 0,25 mm (siehe Punkt 8.6) bedeutet das, daß der Abstand zwischen Pad und Leiter > 0,5 mm sein muß.

8.7 Coverage of Conductors and Filling of Conductor Spaces

a. A conductor running adjacent to a pad shall be fully covered by the mask, provided the nominal distance a between the mask aperture and the conductor is greater than 0.25 mm. In the case of a mask clearance of 0.25 mm, as stated in Item 8.6, the distance between the solder pad and the conductor shall be greater than 0.5 mm.

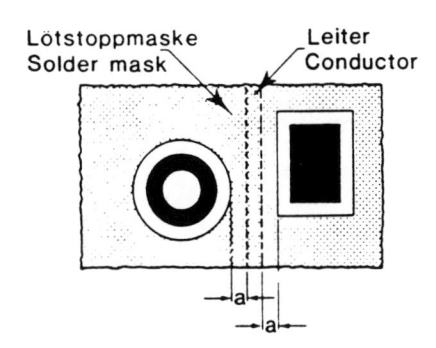

b. Bei Siebdruckmasken besteht das Risiko, daß ein zwischen 2 IC-Pads (mit 2,54 mm Rast) durchgeführter Leiter in Padhöhe nicht vollständig abgedeckt wird. Bei einer Maskenfreifläche von 0,25 mm, wie im Abschnitt 8.6 beschrieben, wird unter folgender Bedingung die Leiterflanke vollständig abgedeckt:

$$b \leq 1{,}54 - d_y \text{ mm}$$

Dabei bedeutet b die Leiterbreite und d_y den Außendurchmesser des Pads.

b. Screen-printed masks incur the risk that a conductor running between two IC pads (on a 2.54 mm grid) does not achieve full edge coverage on a level with the pads. In case of a mask clearance of 0.25 mm, as stated in Item 8.6, full edge coverage implies that the following requirements be met:

$$b \leq 1.54 - d_y \text{ mm},$$

where b indicates the conductor width and d_y indicates the outer diameter of the solder pad.

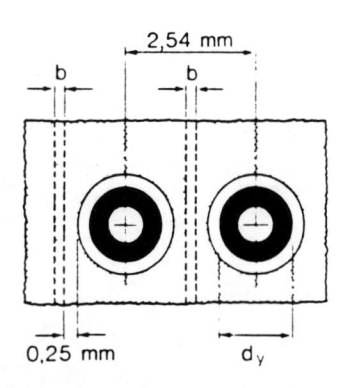

c. The mask shall completely fill the space (insulation distance) between parallel conductors provided the pattern heigth, expressed by the thickness of the copper foil, and the width of the spacing fulfil the following requirements:

Thickness of Copper Foil, t	Width of Spacing, b
$t \leq 17.5$ µm	$b \geq 0.20$ mm
$t = 35$ µm	$b \geq 0.30$ mm
$t = 70$ µm	$b \geq 0.50$ mm

Note
It is implied that the thickness of the copper plating is 25 µm, cf. Item 2.2 b.

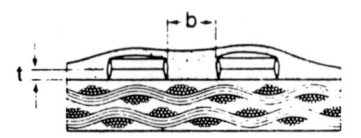

9. Photopolymer Masks

9.1 Definition

In the case of surface mounted boards or mixed technology boards, the concepts solder side, component side, solder mask and insulation mask are no longer unambiguous. Therefore, the specifications below do not distinguish between the said concepts, the only designation being "mask". The location of masks appear from the master drawing.

In the case of hole mounted boards the above concepts are still valid although they are not directly stated in the specifications below.

9.2 Extent

If specified, photopolymer masks are applied as:

- Masks on one or both sides of the board.

9.3 Allgemeine Qualitätsanforderungen

a. Die Maske muß gut auf der Leiterplatte haften und darf nicht durch die beim Maschinenlöten auftretende Wärme beschädigt werden.

   Blankes/oxidiertes Kupfer unter der Maske
   Die Maske darf sich nicht kräuseln, ablösen oder reißen. Die Haftfestigkeit ist nach IPC-TM-650, Punkt 2.4.28.1 zu prüfen. Alternativ kann das Klebeband Typ 691 von 3M Company verwendet werden.

   Maskendruck auf Zinn/Blei
   Für Masken auf Zinn/Blei-Oberflächen gilt folgendes: Ein leichtes Kräuseln der Maske nach dem Maschinenlöten kann akzeptiert werden. Die Maske muß sich der Struktur des erstarrten Zinn/Bleis anpassen. Wenn die Leiter schmaler als 1,27 mm sind, darf kein Ablösen auftreten. Dies sollte nach IPC-TM-650, Punkt 2.4.28.1 geprüft werden. Alternativ kann Klebeband Typ 691 von 3M Company verwendet werden.

b. Die Maske darf keine Risse, z.B. auf der Leiteroberfläche, oder unerwünschte Löcher wie Nadellöcher aufweisen.

c. Es dürfen sich keine Maskenrückstände oder unerwünschter Maskendruck auf den Pads oder in den Bohrungen befinden. Dabei wird für Trockenfilmmasken ein Bohrungsnenndurchmesser ≥ 0,6 mm vorausgesetzt.

   Designhinweis
   Durchverbindungen < 0,6 mm sollten bei Verwendung von Trockenfilmmasken überdeckt werden.

d. Die Maske muß resistent gegen gebräuchliche Lösungsmittel sein: gegen chlorierte und fluorierte Kohlenwasserstoffe, Isopropanol und ähnliches, sowie gegen allgemein übliche Fluxmittel.

e. Unerwünschte Metallpartikel wie z.B. Bohrspäne dürfen weder in der Maske eingeschlossen sein noch auf der Oberfläche haften.

f. Durch ionische oder andere Verunreinigungen, die unter der Maske vorhanden sein könnten, darf keine wesentliche Verschlechterung der elektrischen Eigenschaften der Leiterplatte eintreten. Der Isolationswiderstand zwischen zwei beliebigen Knotenpunkten muß bei einer Prüfspannung von 100 VDC mindestens 500 MΩ betragen.

9.3 General Requirements

a. The mask shall adhere well to the board and shall not be damaged by heat when the board is mass soldered as specified in Item 13.2.

   Bare/Oxidized Copper under the Mask
   No wrinkling, crazing or flaking of the mask shall be found. The adhesion shall be tested according to IPC-TM-650, Item 2.4.28.1. An alternative tape is type 691 from the 3M Company.

   Tin/Lead under the Mask
   A slight wrinkling of the solder mask after mass soldering will be accepted. The mask shall follow the structure of the solidified tin/lead. No flaking shall occur where the underlying conductors are narrower than 1.27 mm. This shall be tested according to IPC-TM-650, Item 2.4.28.1. An alternative tape is type 691 from the 3M Company.

b. The mask shall show no cracks, e.g., on top of conductors, or unintentional holes, e.g., pinholes.

c. Remnants of the mask or unintentional mask on the solder pads or in the plated-through holes/via holes shall not be found. A nominal hole diameter ≥ 0.6 mm is preassumed valid for dry-film masks.

   Design Note
   Via holes < 0.6 mm should be tented in the case of dry film solder masks.

d. The mask shall be resistant to solvents generally used: chlorinated and fluorinated hydrocarbon, isopropanol and the like, and also commonly used fluxes.

e. Unwanted metal particles, e.g., drill chips, shall not be found encapsulated under the mask or adhering to the surface of the mask.

f. Neither ion-contamination nor other contamination shall be found under the mask to such a degree that an essential reduction of the board's electrical characteristics occurs. The insulation resistance between two arbitrary nodal points shall be ≥ 500 MΩ when applying a test voltage of 100 V dc.

g. The mask shall be fully cured. Occurrence of webbing after mass soldering is not acceptable if caused by incomplete curing. Webbing, however, can also be caused by inadequate soldering parameters, e.g., incorrect fluxing, for which reason the user should clarify the reason, e.g., by soldering boards from other deliveries.

h. No delamination, i.e., air pockets under the mask, shall be found.

Design Note

In order to avoid delamination during machining, the distance between the board edges and applied mask edges shall normally not be less than 0.5 mm. This applies also to internal cut-outs.

i. Removal of incorrectly applied masks shall not result in a porous surface of the board or in full or partial exposure of the glass weave.

9.4 Materials

Photopolymer dry-film or liquid-film.

9.5 Thickness of Mask and Conductor Spacing

a. Photopolymer dry-film: Mask thickness of 50, 75 or 100 µm. The relation between the mask thickness, the pattern height expressed by the thickness of the copper foil, and the width of the spacing between parallel conductors appears from the table below, complete filling of the spacing being implied.

Thickness of Copper Foil, t	Thickness of Mask, T	Width of Spacing, b
$t \leq 9$ µm	$T = 50$ µm	$b \geq 0.15$ mm
$t = 17.5$ µm	$T = 75$ µm	$b \geq 0.20$ mm
$t = 35$ µm	$T = 100$ µm	$b \geq 0.30$ mm

Note

It is implied that the thickness of the copper plating is 25 µm, cf. Item 2.2 b.

b. Photopolymer liquid-film: Thickness of mask ≥ 10 μm at the middle of a conductor.

Filling of the spacing between parallel conductors implies the following relation between the thickness of the copper foil and the width of the spacing:

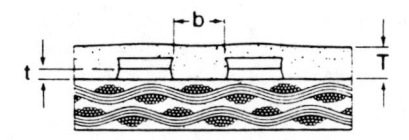

Thickness of Copper Foil, t	Width of Spacing, b
t ≤ 9 μm	b ≥ 0.15 mm
t = 17.5 μm	b ≥ 0.20 mm
t = 35 μm	b ≥ 0.30 mm

Note

Spacing widths down to 0.10 mm implies another technique of application than the serigraphic one, e.g. curtain coating.

9.6 Overlapping of Solder Pads

For technical reasons, the customer's documentation will frequently be based on the same size of solder mask apertures as the solder pads, i.e., the clearance is zero. This can cause the mask to creep 25 to 50 μm over the edges of the solder pads which may create problems when soldering SMT boards.

When the customer has given the PCB manufacturer a general permission to modify the mask documentation, the PCB manufacturer shall modify all mask documentations to increase the clearance from 0 to 0.1 mm. This implies that all mask apertures exhibit a clearance of zero.

When no general permission to modify the mask documentation exists, and technical production considerations make a modification desirable, the customer's approval in the individual cases is implied.

Kohlepastendruck	Carbon Ink Printing
Wenn ein Abstand zwischen Pad und Maske von 0 mm gefordert wird und eine Filmänderung nicht zulässig ist, dann ist eine Überdeckung des Padrandes von 0,1 mm zulässig.	If a modification of the mask documentation is not permitted, and the clearance is zero, an overlapping of the pad rim by 0.1 mm is accepted.

9.7 Leiterabdeckung / Coverage of Conductors

Ein an einem Pad vorbeiführender Leiter muß vollständig abgedeckt sein, wenn der Abstand a zwischen Leiterkante und Maskenausschnitt mehr als 0,1 mm beträgt.

A conductor adjacent to a solder pad, shall be covered by the mask provided the nominal distance a between the mask aperture and the conductor is greater than 0.1 mm.

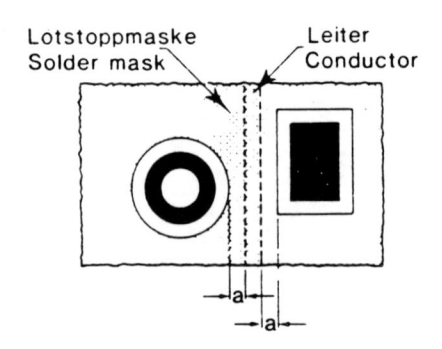

Lotstoppmaske / Solder mask — Leiter / Conductor

10. Kohlepastendruck / Carbon Ink Printing

10.1 Anwendung / Application

Kohlepasten werden im Siebdruck als Tastaturkontakte sowie LCD-Kontakte und Streifen aufgebracht.

Carbon ink printing is used for screen-printed keyboard contacts, LCD contacts and straps.

10.2 Vorbedingung / Preconditions

Die im folgenden angeführten Anforderungen setzen eine Dicke der Kupferfolie von

$$t \leq 17.5\,\mu m$$

voraus.

The requirements stated below imply the following thickness t of the laminate's copper foil:

$$t \leq 17.5\,\mu m$$

10.3 Allgemeine Anforderungen / General Requirements

a. Die Oberfläche gedruckter Kontakte und Verbindungen muß eben, kratzerfrei und sauber sein und ein gleichmäßiges Aussehen von Kontakt zu Kontakt bzw. Verbindung zu Verbindung aufweisen.

a. The surface of printed contacts and straps (jumpers) shall be smooth, with no scratches, clean and with an uniform appearance from contact to contact, alternatively from strap to strap.

Kohlepastendruck	Carbon Ink Printing
b. Auf den Lötflächen darf sich keine ungewollte, vom Siebdruckprozeß herrührende Kohleschicht (Kohlefilm) befinden.	b. Unintentional carbon ink (carbon film) originating from the screen-printing process shall not be found on the solder pads of the board.
c. Die Kontaktoberfläche muß gegen allgemein übliche Lösungsmittel beständig sein: gegen chlorierte und fluorierte Kohlenwasserstoffe, Isopropanol und ähnliches sowie gegen allgemein übliche Fluxmittel.	c. The contact surface shall be resistant to solvents generally used: chlorinated and fluorinated hydrocarbon, isopropanol and the like, and also commonly used fluxes.

10.4 Materialien

Es wird mit einer fein- oder grobkörnigen Kohlepaste gedruckt.

10.4 Materials

The printing is performed with conductive carbon ink which can be fine-grained or course-grained.

10.5 Detaillierungsgrad des Leiterbildes beim Siebdruck mit Kohlepaste

Da die Druckgenauigkeit von der Korngröße der Kohlepaste abhängt, muß der Leiterplattenhersteller den Kohlepastentyp entsprechend dem Detaillierungsgrad des Leitermusters auswählen, es sei denn, daß der Kunde eine bestimmte Paste vorgeschrieben hat.

Der Detaillierungsgrad wird durch die nachstehend aufgeführten Parameter bestimmt. Die für das Design maßgebenden Werte sind in Klammern angegeben.

M: Druckbreite des Kohleleiterbildes
($M \geq 0,3$ mm)

N: Isolationsabstand Kohle/Kohle
($N \geq 0,3$ mm)

P: Isolationsabstand Kohle/Kupfer
($P \geq 0,5$ mm)*

Q: Überlappung Kohle/Kupfer
($Q \geq 0,3$ mm)

R: Abstand Kohle/Lötstoppmaske
($R \geq 0,3$ mm)*

* Die Werte gelten für Fotopolymermasken. Für siebgedruckte Masken erhöht sich der Wert um 0,2 mm.

10.5 Degree of Carbon Pattern Detail

Since the printing accuracy depends upon the grain size of the carbon ink, the PCB manufacturer must choose the type of carbon ink in accordance with the degree of detail of the carbon pattern, if the type of carbon ink has not been specified by the customer.

The degree of detail is determined by the parameters stated below. Design values are stated in parentheses. See the figures below.

M: Printing width of carbon pattern
($M \geq 0.3$ mm)

N: Carbon/carbon insulation distance
($N \geq 0.3$ mm)

P: Carbon/copper insulation distance
($P \geq 0.5$ mm)*

Q: Carbon/copper overlapping
($Q \geq 0.3$ mm)

R: Carbon/solder mask clearance
($R \geq 0.3$ mm)*

* Valid for photopolymer masks. In the case of screen-printed masks, the value is increased by 0.2 mm.

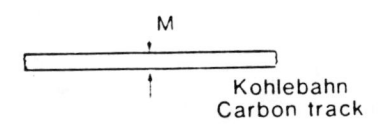

M
Kohlebahn
Carbon track

Kohlepastendruck | Carbon Ink Printing

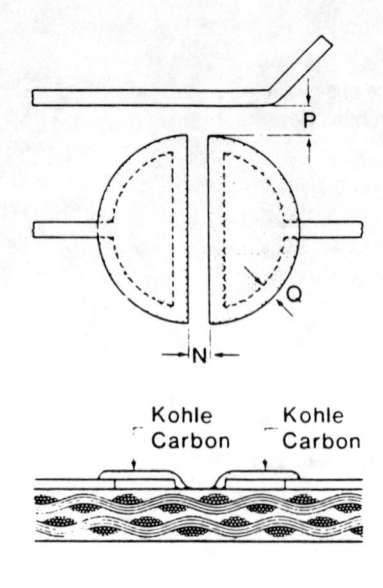

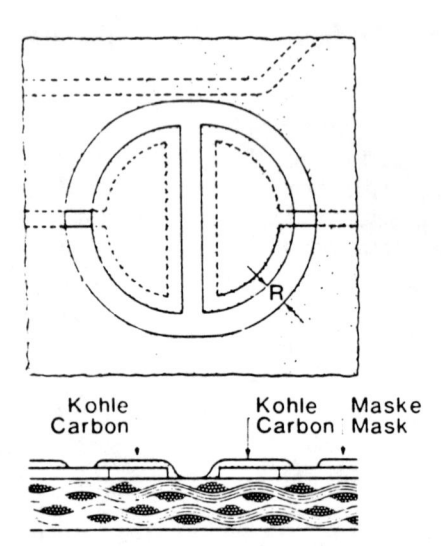

10.6 Allgemeine Anforderungen an das Leiterbild (75 %-Regel)

Bei allen willkürlichen Kombinationen der in den Abschnitten 10.7 bis 10.10 beschriebenen Anforderungen bezüglich des Kohledruckbildes sind die folgenden grundsätzlichen Punkte a bis c zu erfüllen.

a. Die tatsächliche Druckbreite eines Kohleelementes darf 75 % des Nennwertes nicht unterschreiten.*

b. Der Isolationsabstand N zwischen benachbarten Kohleflächen darf 75 % des Nennabstandes nicht unterschreiten.*

c. Der Isolationsabstand zwischen Kohle und Kupfer darf 75 % des Nennwertes* nicht unterschreiten, wenn dieser vorher um die Druckungenauigkeit, d.h. den Versatz ΔF korrigiert wurde, d.h.

10.6 General Requirements on Pattern (75 % Rule)

With any arbitrary combination of the requirements stated in Items 10.7 to 10.10 concerning the carbon pattern, the following primary requirements a to c shall be fulfilled.

a. The effective printing width M of a carbon element shall not be reduced below 75 % of the nominal value.*

b. The carbon/carbon insulation distance N shall not be reduced below 75 % of the nominal value.*

c. The carbon/copper insulation distance P shall not be reduced below 75 % of the nominal value*, when, in advance, the latter has been corrected for the printing inaccuracy = the displacement ΔF, i.e.,

75 % of (P* - ΔF).

The displacement ΔF appears from Item 10.13.

Design Note
In the case of a pattern where the above requirements cannot be fulfilled, an extra layer of insulation mask can be applied at the critical areas, cf. Item 10.18.

- References: Customer's documentation, cf. Item 4.1.

10.7 Edge Definition of Carbon Pattern

The unevenness U (edge definition) of the carbon pattern edges shall be ≤ 0.2 mm, crest to trough.

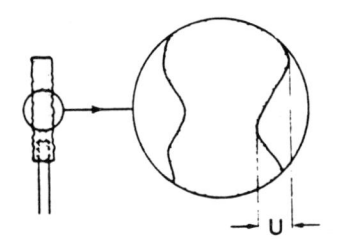

10.8 Indentations and Projections

A few unwanted indentations and projections along the carbon pattern edges are accepted under the following conditions:

a. The depth D of an indentation shall be less than 1 mm.

b. The size H of a projection shall be less than 1 mm.

Note
The 75 % rule (Item 10.6) implies that indentations or projections of 1 mm are accepted only in the case of printing widths, alternatively insulation distances, of 4 mm and above

c. The length L of an indentation or a projection shall be less than 1 mm.

| Kohlepastendruck | Carbon Ink Printing |

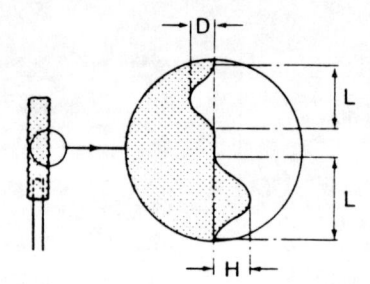

d. Die durchschnittliche Fehlerzahl (Vorsprünge und Einschnürungen) entlang eines Leiterrandes muß weniger als 1 Fehler pro 50 mm Kantenlänge betragen.

10.9 Kohlerückstände

Unerwünschte festhaftende Kohlerückstände, die beim Siebdruck entstanden sein können, werden ebenso wie unerwünschte Metallpartikel betrachtet, s. Punkt 4.7. Sie sind unter folgenden Bedingungen zulässig:

a. Die größte Abmessung darf 1 mm sein.
b. Der Abstand zwischen den Kohleflecken muß mindestens 50 mm betragen.
c. Wenn sich Kohlerückstände zwischen leitenden Teilen des Leiterbildes befinden (Leiter, Pads und Kohlebauelemente), darf die unter den Punkten 4.2 c und 10.6 b und c beschriebene 75 %-Regel nicht verletzt werden.

10.10 Fehlstellen in der Kohleschicht

Unerwünschte Fehlstellen in der Kohleschicht sind unter folgenden Voraussetzungen zulässig:

a. In einer auf Kupfer gedruckten Kohleschicht sind keine Fehlstellen zulässig.
b. Wenn die Kohleschicht direkt auf das Basismaterial gedruckt wird, können Fehlstellen in der Kohleschicht unter folgenden Bedingungen akzeptiert werden:
  - Der Lochdurchmesser darf maximal 0,5 mm sein.
  - Es dürfen nur weniger als 5 Fehlstellen pro cm^2 Kohlefläche vorkommen.

d. The average number of defects (projections and indentations) along a pattern edge shall be less than one defect per 50 mm of edge length.

10.9 Carbon Specks

Unwanted attached carbon specks caused by accidental splashes due to screen-printing the ink, are regarded on a par with unintentional metal particles, cf. Item 4.7. They are accepted under the following conditions:

a. The largest dimension to be 1 mm.
b. The mutual distance to be greater than 50 mm.
c. Where carbon specks are located between conductive pattern elements (conductors, solder pads and carbon elements), the 75 % rule of Items 4.2 c and 10.6 b and c shall not be violated.

10.10 Voids in the Carbon Layer

Unwanted voids in the carbon layer are accepted under the following conditions:

a. No voids shall be found in the carbon layer where the carbon pattern is printed on top of copper.
b. Where the carbon pattern is printed on bare base laminate, voids in the carbon layer are accepted under the following conditions:
  - the diameter of the voids is max. 0.5 mm
  - the number of voids shall be less than 5 voids per cm^2 of carbon area.

10.11 Carbon Overlapping

Uncovered conductors terminating carbon elements shall not be found when the following requirements are fulfilled:

a. Carbon/copper overlapping $Q \geq 0.3$ mm

b. Carbon/solder mask overlapping $S \geq 0.4$ mm in the case of photopolymer masks and $S \geq 0.5$ mm for screen-printed masks.

Design Note

Execution A, cf. the figure below, implies solder coating and hot-air levelling in order to avoid exposed copper. If so, the carbon layer shall not be protected by a peelable mask, cf. Item 10.16 b.

Execution B is used in the case of bare-copper boards, provided the carbon printing proceeds a little in over the mask and thereby seals the intermediary part of the conductor.

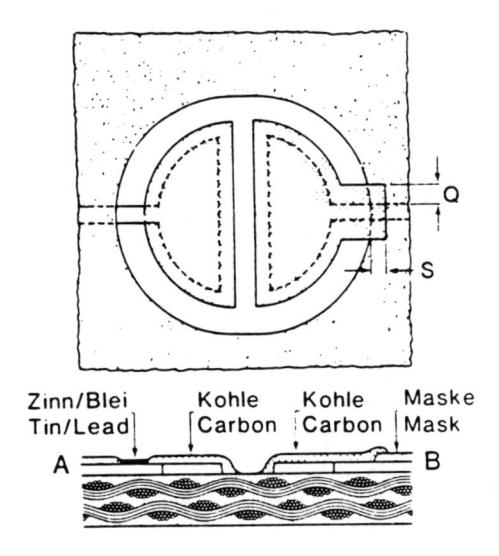

10.12 Termination Area

The accuracy of the ink printing can be increased by printing on a bare base laminate. If so, the termination areas (carbon/copper interface) should be located out-side the effective contact areas EK.

| Kohlepastendruck | Carbon Ink Printing |

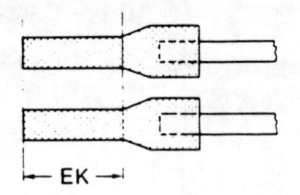

Die Anschlußlänge T* soll mindestens 1,0 mm betragen und darf durch eine ungenaue Registrierung nicht unter 0,7 mm absinken.

The length T* of the termination areas shall be min. 1.0 mm and shall not be reduced below 0.7 mm as a result of misregistered ink printing.

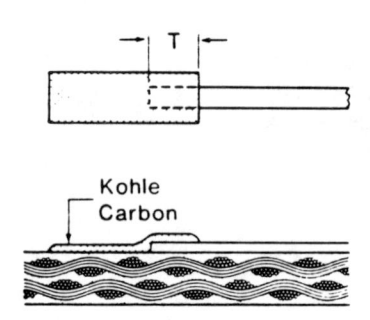

Wenn eine Isoliermaske aufgebracht wird, so darf diese den Anschlußbereich Kupfer/Kohle nur soweit überdecken, daß mindestens noch eine Anschlußüberlappung T_1 = 0,5 mm bestehen bleibt (s. Skizze).

When an insulation mask is applied, it shall not cover more of the termination areas than corresponding to a maximum reduction of T (cf. the figure below) to T_1 = 0.5 mm.

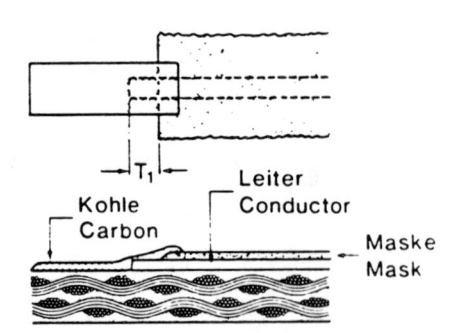

Wenn für das Kohleleiterbild ein höherer Detaillierungsgrad, als in Abb. 10.5 beschrieben, gefordert wird, dann sind die entsprechenden Details zwischen Kunde und Leiterplattenhersteller zu vereinbaren.

If the degree of detail of the carbon pattern is higher than stated in Item 10.5, details of the design shall be determined in collaboration with the PCB manufacturer.

* Referenz: Fertigungsunterlagen des Kunden, s. Abschnitt 4.1.

* Reference: Customer's documentation, cf. Item 4.1.

| Kohlepastendruck | Carbon Ink Printing |

10.13 Registrierung des Kohleleiterbildes

Die Ausrichtung des Siebes für den Kohledruck und Kontrollmessungen des Kohleleiterbildes werden anhand von Registriermarken im Kupferleiterbild, die dicht am Kohlebereich liegen, vorgenommen. Beispiele hierfür mit siebgedruckten Registriermarken zeigt die nachstehende Skizze.

10.13 Registration of Carbon Pattern

Screen-printing of the carbon ink and measurement of the registration of the carbon pattern are performed on the basis of selected reference marks in the copper pattern, located adjacent to the carbon area. Examples of targets with screen-printed check marks in correct registration are shown below.

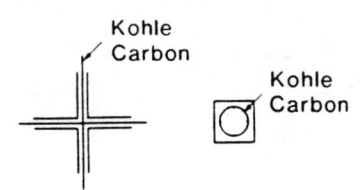

```
        Kohle
       / Carbon              Kohle
  ──┼──              ▫        Carbon
        │
```

Der Versatz des Kohleleiterbildes in Bezug auf die vorstehend beschriebenen Referenzmarken darf für Leiterplatten bis 200 x 200 mm maximal 0,2 mm in jeder Richtung betragen. Bei größeren Leiterplatten werden je angefangene 100 mm Kantenlänge über 200 mm mindestens 0,05 mm hinzuaddiert. Das gilt sowohl für einzelne als auch für im Nutzen addierte Leiterplatten bis zu einer Größe von 450 x 450 mm.

It is required that the displacement ΔF of the carbon pattern in any direction with respect to the reference marks mentioned above be less than 0.20 mm for board sizes up to 200 x 200 mm. In the case of larger boards, 0.05 mm is added for every commenced 100 mm of length beyond 200 mm. This applies to individual boards, alternatively boards delivered in panels, up to 450 x 450 mm.

10.14 Haftfestigkeit

Der Kohledruck muß gut auf dem darunterliegenden Material haften und darf bei nachfolgenden Wärmebelastungen (z.B. beim Tauchverzinnen, Hot Air Levelling, Tempern, Maschinenlöten) nicht beschädigt werden, s. Abschnitt 13.2.

Die Haftfestigkeit ist nach IPC-TM-650 Punkt 2.4.28 zu prüfen. Die Kohleschicht darf weder aufplatzen, reißen, delaminieren noch sich ablösen.

10.14 Adhesion

The carbon printing shall adhere well to the underlying material and shall not be damaged by heat, partly during the PCB manufacturer's solder coating and hot-air levelling process of the board, partly when the board is prebaked and mass soldered. cf. Item 13.2

The adhesion shall be tested according to IPC-TM-650, Item 2.4.28. No cracking, delamination or flaking is accepted.

10.15 Widerstand und Isolationswiderstand

Um sicherzustellen, daß die spezifizierten Werte für den Kohlewiderstand und den Isolationswiderstand zwischen den Kohleelementen eingehalten werden, kann der Leiterplattenhersteller diese Werte prüfen. Die Prüfung kann an einem ausgewählten Testbereich oder einer außerhalb der Leiterplatte aufgedruckten Testgeometrie erfolgen. Beispiele für Testgeometrien sind unten dargestellt.

10.15 Resistance/Insulation Resistance

In order to ensure that specified values of carbon resistance and insulation resistance of carbon elements are met, a test can be performed by the PCB manufacturer. The test is based on a selected test area or a special test pattern. The latter can be located by the PCB manufacturer just outside the board area. Examples of test patterns are shown below.

| Kohlepastendruck | Carbon Ink Printing |

a. Der Kohlewiderstand wird mit einem an die Punkte A und B angeschlossenen Ohmmeter gemessen. Der gemessene Widerstand soll so gering wie möglich sein und darf nach einem Test unter Umweltbedingungen, s. Punkt 10.20, einen Wert von 200 Ω nicht überschreiten.

a. The carbon resistance is measured by means of a megohmmeter connected to terminals A and B. The resistance value measured shall be as low as possible and shall not exceed 200 Ω after an environmental testing as per Item 10.20.

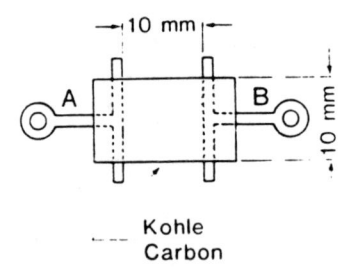

Kohle
Carbon

b. Die Kontaktoberfläche muß leitend sein. Wenn der Kontaktwiderstand zu externen Kontaktelementen, z.B. einer leitenden Gummimatte oder Kontaktfedern, spezifiziert ist, muß der Kunde die Prüfmethode vorschreiben und entsprechende Prüfeinrichtungen zur Verfügung stellen.

c. Der Isolationswiderstand zwischen den Kohleelementen wird mit einem Hochspannungsohmmeter gemessen, das an die Punkte A und B angeschlossen wird.

Wenn Prüfspannungs- und Widerstandswerte nicht spezifiziert sind, ist die Prüfung mit 500 VDC durchzuführen, wobei der gemessene Widerstand 500 MΩ betragen soll.

b. The surface of the contacts shall be conductive. When the contact resistance to external contact elements, e.g. a conductive rubber mat or contact springs, is specified, the customer shall prescribe the test method and deliver test equipment.

c. The insulation resistance between carbon elements is measured by means of a high-voltage megohmmeter connected to carbon elements A and B.

In case no test voltage and resistance values are specified, the test shall be performed at a voltage of 500 V dc, and the resistance measured shall be above 500 MΩ.

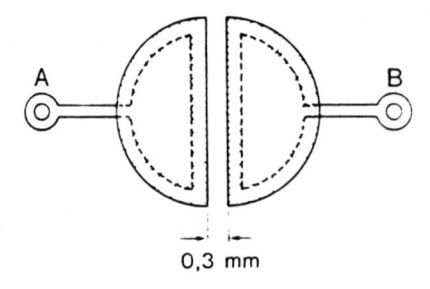

d. Wo ein Kohlestreifen über eine Kupferfläche oder einen Leiter mit dazwischenliegender Isolierschicht gedruckt ist, wird der Isolationswiderstand mit einem Ohmmeter, das an die Kontakte A und C oder B und C angelegt wird, gemessen.

d. Where a carbon strap is printed across a copper area or across a conductor with an intermediate insulation mask, the insulation resistance is measured by means of a high-voltage megohmmeter connected to terminals A and C, or B and C.

| Kohlepastendruck | Carbon Ink Printing |

Wenn keine Prüfspannung und Widerstandswerte angegeben sind, ist die Prüfung mit 500 VDC auszuführen, wobei der gemessene Widerstand über 100 MΩ betragen soll.

If no test voltage and resistance values are specified, the test shall be performed at a voltage of 500 V dc, and the resistance measured shall be above 100 MΩ.

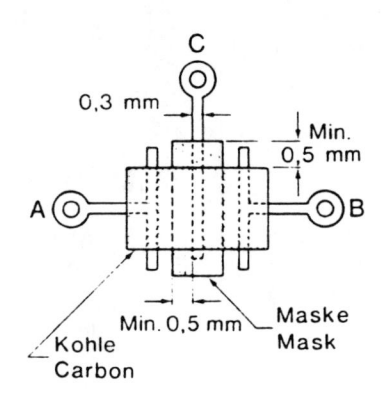

Diagram: Rectangular pad with terminals A (left), B (right), C (top). Dimensions shown: 0,3 mm, Min. 0,5 mm (vertical), Min. 0,5 mm (horizontal). Labels: Kohle/Carbon, Maske/Mask.

10.16 Oberflächenschutz (Abziehbare Maske)

Es kann verschiedene Gründe für das Aufbringen einer temporären Maske (abziehbaren Maske) zum Schutz von Kohlekontakten geben.

a. Die Leiterplatte soll durch einen Lötlack oberflächengeschützt werden. In diesem Fall muß der Leiterplattenhersteller dafür Sorge tragen, daß eine abziehbare Maske aufgetragen wird, es sei denn, der Kunde habe eine Lackierung ohne vorausgehenden Schutz der Kohlekontakte spezifiziert.

b. Der Kunde läßt nicht zu, daß die Leiterplatte ohne Schutz der Kohlekontakte heißluftverzinnt wird. In diesem Fall spezifiziert der Kunde die temporäre Maske.

c. Die Kohlekontakte befinden sich auf der Lötseite der Leiterplatte und der Kunde will die Leiterplatte nach dem Maschinenlöten nicht reinigen. Der Kunde spezifiziert die Aufbringung einer temporären Maske.

Hinweis
Zwischen der temporären Maske und der Kohlepaste muß Kompatibilität bestehen, da die letztere sonst geschädigt werden kann. Wenn der Kunde selbst die abziehbare Maske aufträgt, muß er sie in den Fertigungsunterlagen spezifizieren oder mit dem Leiterplattenhersteller abstimmen.

10.16 Surface Protection (Peelable Mask)

There may be several reasons to protect the surface of carbon contacts with a temporary mask (peelable mask).

a. The board is to be surface-protected by solder lacquer. If so, the PCB manufacturer shall make sure to apply a peelable mask, unless the customer has specified lacquering without any preceding protection of the carbon contacts.

b. The customer cannot accept that the board is solder coated and hot-air levelled without protection of the carbon contacts. The customer specifies the application of a peelable mask.

c. The carbon contacts are located on the solder side of the board, and the customer does not wish to clean the board after mass soldering. The customer specifies the application of a peelable mask.

Note
It is required that there be compatibility of materials between the peelable mask and the carbon ink since the latter otherwise may be damaged. If the customer himself applies the peelable mask, the type shall be specified in the PCB documentation or be agreed on with the PCB manufacturer.

| Kohlepastendruck | Carbon Ink Printing |

10.17 Ausführung

a. Der Siebdruck darf nur auf einer sauberen Oberfläche erfolgen, sei es das Basismaterial, Kupfer oder eine Isoliermaske.

b. Die Kohlepaste muß vollkommen ausgehärtet sein, so daß keine schädlichen Rückstände an flüchtigen Bestandteilen in der Kohleschicht zurückbleiben.

10.18 Leiter unter der Kohleschicht

Wenn die Kohleschicht über einen Leiter gedruckt wird, z.B. als Verbindungselement, dann muß der Leiter mit einer Isolierschicht abgedeckt sein, die die in Abschnitt 10.25 angegebenen Anforderungen erfüllt. Die Isoliermaske kann mit einer zusätzlichen Isoliermaske überdruckt werden, eventuell nur in den kritischen Bereichen.

Hinweis
Zwischen dem Maskenmaterial und der Kohlepaste muß Verträglichkeit bestehen, da diese sonst beschädigt werden könnte.

Designhinweis
Um eine zuverlässige elektrische Verbindung zwischen der Kohle und der Anschlußfläche sicherzustellen, muß die Anschlußstelle eben und flach sein. Trockenfilmmasken eignen sich nicht besonders gut, da sie hohe, steile Flanken aufweisen. Besser eignen sich Flüssigresiste, da sie niedrige Ränder bilden. Mit siebgedruckten Masken läßt sich eine gewisse Nivellierung erreichen, besonders wenn man bei einem doppelten Druck die zweite Maske um beispielsweise 0,20 mm zurücksetzt, siehe Stelle "A" in untenstehender Skizze.

Im Gegensatz dazu ist der Übergang "B" steiler und höher, da sich beide Maskenlagen genau decken.

10.17 Execution

a. Screen-printing of carbon ink shall be performed on a clean substrate only, whether this be the base laminate, the copper or the insulation mask.

b. The carbon ink shall be fully cured so no harmful remnants of volatile components remain in the carbon element.

10.18 Conductors under Carbon

Where the carbon ink is printed across a conductor, e.g. as a strap (jumper), the conductor shall be covered with an insulation mask tight enough to fulfil the requirement for insulation resistance stated in Item 10.15 d. The insulation mask applied can possibly be overprinted with an extra insulation mask, possibly at the critical areas only.

Note
It is required that there be compatibility of materials between the mask material and the carbon ink since the latter otherwise may be damaged.

Design Note
In order to ensure a reliable electrical connection between the carbon printing and the termination area, the interface should be even and low. Dry film masks form high edges for which reason they are not very suitable. Liquid film masks form low edges and are therefore more suited. Screen-printed masks give a certain levelling, particularly when using a double printing with a slight withdrawal, e.g., 0.20 mm, as shown at A below.

The contrast is shown at B where the transition is steeper and higher than at A because of a coinciding mask printing.

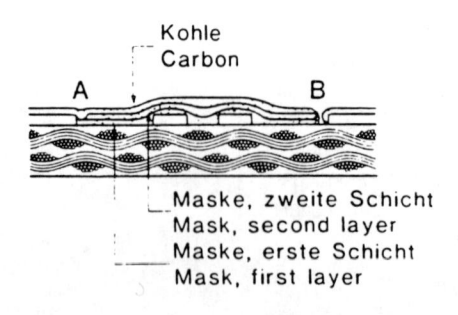

10.19 Auswirkung des Lötprozesses

a. Die Kohleelemente müssen die in Abs. 13.2 a beschriebenen Lötbedingungen ohne Schädigung aushalten.

b. Die Kohleelemente müssen resistent sein gegen Tauchverzinnung, Hot Air Levelling und die zugehörigen Fluxmittel. Eventuell kann hierbei eine abziehbare Maske als Oberflächenschutz verwendet werden, siehe Absatz 11.

c. Beim Löten dürfen weder Risse noch Blasen in der Kohleschicht noch eine Delaminierung zwischen Kohleschicht und darunterliegendem Material auftreten.

d. Es dürfen sich auf der Kohleschicht keine Netze (Zinnfäden) bilden.

10.20 Prüfung unter Umweltbedingungen

Alle vorstehenden Anforderungen sind sowohl vor als auch nach einer Prüfung gemäß IEC 68-2-30, Test Db, + 40 °C und zwei Zyklen, zu erfüllen. Die Anforderungen müssen ebenfalls nach einer 56-Tage-Lagerung bei 85 °C (trockene Wärme) erfüllt werden.

Wenn die Leiterplattenspezifikation einen Salznebeltest vorschreibt, dann müssen die genannten Anforderungen nach Durchführung dieses Tests erfüllt werden.

11. Abziehbare Masken

11.1 Anwendungsbereich

Abziehbare Masken dienen zum Abdecken bestimmter Bohrungen in Leiterplatten, damit beim Maschinenlöten kein Lot eindringt oder zum Schutz von Kohleelementen und Goldkontakten während des Maschinenlötens.

11.2 Ausführung

Abziehbare Masken werden im Siebdruck aufgebracht.

11.3 Allgemeine Anforderungen

a. Die abziehbare Maske muß so gut auf der Leiterplatte haften, daß sie bei sachgerechter Handhabung nicht abfällt.

10.19 Effect of Soldering

a. The carbon elements shall resist the soldering conditions stated in Item 13.2 a.

b. The carbon elements shall also resist solder coating and hot-air levelling, including the flux used, possibly with a surface protection in the form of a peelable mask. Cf. Item 11.

c. Soldering shall not result in the formation of cracks or blisters in the carbon printing, or delamination between the carbon elements and the underlying material.

d. No webbing shall be found on the carbon elements.

10.20 Environmental Testing

All preceding requirements shall be fulfilled before as well as after an evironmental testing as per IEC 68-2-30, Test Db, + 40 °C and 2 cycles. The requirements shall also be fulfilled after storage for 56 days at 85 °C (dry heat).

If the PCB documentation also states a salt mist testing, the preceding requirements shall be fulfilled after performing the test.

11. Peelable Masks

11.1 Application

Peelable masks are used in order to protect selected solder holes against closing during mass soldering or to protect carbon elements and gold plated contacts during mass soldering.

11.2 Execution

Peelable masks are applied by screen-printing.

11.3 General Requirements

a. The peelable mask shall adhere so well to the underlying material that it does not fall off with ordinary handling of the board.

Abziehbare Masken	Peelable Masks
b. Die abziehbare Maske muß folgende Belastungen aushalten, ohne sich zu lösen oder nicht entfernbar zu werden: - zweimaliges Erwärmen der Leiterplatte, wie unter 13.2 d beschrieben. **Bemerkung** Da gewisse Masken keinen Temperaturprozeß vertragen sondern austrocknen, sollte der Leiterplattenhersteller konsultiert werden. - Zweimaliges Aufheizen der Leiterplatte beim Maschinenlöten wie unter Punkt 13.2 a spezifiziert. c. Nach dem Löten der Leiterplatte muß sich die abziehbare Maske, vorzugsweise an einem Stück, abziehen lassen. Dabei dürfen keine Maskenrückstände in den Bohrungen oder auf der Leiterplattenoberfläche zurückbleiben. **Bemerkung** Die letzte Anforderung gilt für metallisierte Löcher. Wenn die (gestanzten oder gebohrten) Löcher nicht metallisiert sind, kann nicht erwartet werden, daß keine Rückstände zurückbleiben. d. Die abziehbare Maske muß gegen gebräuchliche Lösungsmittel resistent sein: gegen chlorierte und fluorierte Kohlenwasserstoffe, Isopropanol, übliche Fluxmittel und ähnliches. e. Die abziehbare Maske darf die Eigenschaften der darunterliegenden Kohleschicht nicht so stark verändern, daß die unter 10.15 genannten Werte nicht eingehalten werden können.	b. The peelable mask shall withstand the following conditions without becoming either loose or nonremovable: - A preceding baking of the board as per Item 13.2 d. **Note** Since certain masks do not withstand baking but dry out, the PCB manufacturer should be consulted. - Heating the board twice by mass soldering as specified in Item 13.2 a. c. After soldering the board, it shall be possible to remove the peelable mask, preferably in one piece, and no mask material shall be left in the holes or on the areas it covered. **Note** The last requirement is valid for plated through holes. In the case of nonplated through holes (punched or drilled), a complete fulfilment of the last requirement cannot be expected. d. The peelable mask shall be resistant to solvents generally used: chlorinated and fluorinated hydrocarbon, isopropanol and the like, and also commonly used fluxes. e. The peelable mask shall not change the characteristics of underlying carbon elements, if any, to such a degree that the values stated in Item 10.15 cannot be met.

11.4 Besondere Anforderungen | 11.4 Special Requirements

a. Die Ungleichmäßigkeit U (Kantenschärfe) des Maskenrandes soll von der Spitze bis zum Grund ≤ 0,7 mm sein.	a. The unevenness U (edge definition) of the peelable mask edges shall be < 0.7 mm, crest to trough.

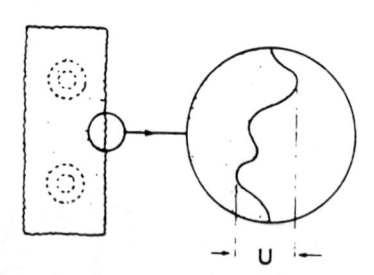

b.	Unter der Voraussetzung, daß eine nominale Überlappung von abziehbarer Maske und Leiterbild von V ≥ 0,7 mm vorgesehen ist, muß das Leiterbild vollständig abgedeckt sein. Ein Überdrucken von Lötaugen darf nicht vorkommen, wenn der Nominalabstand W zwischen Lötaugen und Maskenrand ≥ 0,7 mm ist (s. Skizze).

b. Uncovered pattern (solder pads or contacts) shall not be found when the peelable mask has a nominal overlapping V ≥ 0.7 mm. Overprinting of solder pads with a peelable mask shall not occur when the distance W ≥ 0.7 mm.

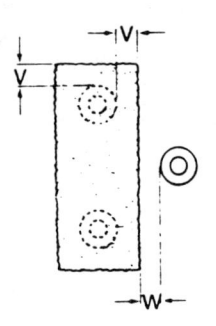

c. Bohrungen mit einem Durchmesser > 1,0 mm brauchen nicht vollständig von der abziehbaren Maske verschlossen zu werden.

d. Beim Auftragen der Maske darf diese nicht durch die abzudeckenden Bohrungen fließen und auf der Gegenseite der Leiterplatte austreten.

Das gilt auch für Bohrungen mit einem Durchmesser D ≥ 3,0 mm, sofern die Maske so konzipiert ist, daß bis maximal 0,3 mm über den Lochrand gedruckt wird (siehe Skizze).

c. It is accepted that holes larger than 1.0 mm are not completely closed by the peelable mask.

d. Application of a peelable mask shall not cause mask ink to flow down through holes to be covered and out on the solder pads on the opposite side.

This requirement is also valid for holes with a diameter D ≥ 3.0 mm, provided the mask is designed in such a way that it projects no more than 0.3 mm beyond the hole edge.

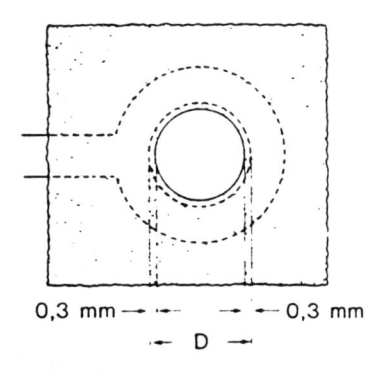

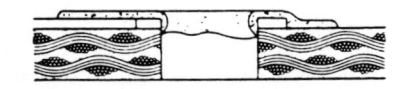

Oberflächenschutz	Protective Coating

e. In der Maske dürfen sich keine ungewollten Fehlstellen mit einem Durchmesser d ≥ 0,5 mm befinden.

12. Oberflächenschutz

Um die Lötbarkeit der Leiterplatte sicherzustellen, soll immer ein Oberflächenschutz aufgebracht werden.

Folgende Arten von Schutzüberzügen, die in den Fertigungsunterlagen spezifiziert werden müssen, können aufgebracht werden:

a. Zinn/Blei-Galvanisierung mit nachfolgendem Aufschmelzen
b. Heißluftverzinnung
c. Lötschutzlack
d. Chemische (stromlose) Verzinnung
e. Andere Arten von Schutzbeschichtungen

12.1 Allgemeine Anforderungen

a. Die Schutzbeschichtung muß die Lötbarkeit innerhalb der Lagerzeit sicherstellen, siehe Abschnitt 13.3.
b. Die Schutzbeschichtung muß auf eine vollständig lötbare Oberfläche aufgebracht werden.
c. Die Bohrungen dürfen nicht durch die Schutzbeschichtung verstopft werden.
d. Die effektive Kontaktfläche von Steckkontakten, Schalterkontakten, Tastaturen und ähnlichem darf nicht von der Schutzbeschichtung überdeckt werden.

12.2 Zinn/Blei-Galvanisierung mit nachfolgendem Aufschmelzen

Außer dem Aufschmelzen benötigen mit Zinn/Blei galvanisierte Leiterplatten keinen weiteren Oberflächenschutz. Die Anforderungen gehen aus Abschnitt 2.3 und 3 hervor. Masken und Komponentendruck sind, falls notwendig, nach dem Aufschmelzen aufzubringen.

12.3 Heißluftverzinnung (Hot Air Levelling)

a In vieler Hinsicht kann die Heißluftverzinnung mit der Zinn/Blei-Galvanisierung mit nachfolgendem Aufschmelzen verglichen werden. Daher gelten die gleichen Anforderungen, die in Abschnitt 2.3 und 3 genannt sind.

---

e. Unintentional voids in the peelable mask with a diameter d ≥ 0.5 mm shall not be found.

12. Protective Coating

A protective coating for teh preservation of the board's solderability shall always be performed.

The following types of protective coating can be employed, and the desired execution will be stated on the master drawing.

a. Tin/lead plating with reflowing
b. Solder (tin/lead) coating and hot-air levelling
c. Lacquering
d. Immersion tinning
e. Other types of protective coating

12.1 General Requirements

a. The protective coating shall be able to ensure the solderability within the storage time. See Item 13.3.
b. The protective coating shall be applied on a fully solderable surface.
c. Solder holes shall not become clogged because of the protective coating.
d. The effective contact area of edge connectors, switch contacts, keyboard switches and the like shall not be covered by the protective coating.

12.2 Tin/Lead Plating with Reflowing

Aside from reflowing, no further surface protection of tin/lead plated boards is to be performed. The requirements appear from Items 2.3 and 3. Masks as well as component notations shall, if any, be applied after performing the reflowing.

12.3 Solder Coating and Hot-Air Levelling

a. In many respects, solder coating and hot-air levelling can be compared with tin/lead plating followed by reflowing. Therefore, the same requirements as stated in Items 2.3 and 3 apply.

b. Unless otherwise stated on the master drawing, masks, if any, shall be applied on bare copper. The copper surface under the masks may appear as oxidized copper in order to achieve a better adhesion or a better visual contrast.

12.4 Protective Lacquering

If, in order to secure an automatic component insertion, no tin/lead in the plated-through holes is desired, the surface protection can be executed as a lacquering.

a. A protective lacquer of a heavily polymerizing type (water-dip lacquer) shall be used only when agreed upon.

b. Visible and excessive accumulations of lacquer (drops) shall not be found in areas to be soldered.

c. Incomplete coverage of the copper surface shall not be found, and the lacquer shall exhibit good adhesion.

12.5 Immersion Tin

By special agreement only, the surface protection may be immersion tin.

12.6 Other Types of Protective Coating

Other types of protective coating, e.g., gold plating, nickel plating, etc., may be applied according to special agreement between the customer and the PCB manufacturer. This applies also to selective gold plating or to electroless nickel used for chip-on-board assembly.

13. Soldering

13.1 General Requirements

a. Full solderability according to IPC-S-804, Item 4.2 (Rotary Dipping Test), is required.

b. The soldering time for a 4-layer board with a thickness of 1.6 mm is 3 sec. In the case of more than 4 layers, the soldering time is increased by 1 sec. for

Ist die Leiterplatte dicker als 1,6 mm, dann ist die Lötzeit um 1 Sekunde pro angefangene 0,8 mm über 1,6 mm hinaus zu verlängern.	everey commenced group of 4 layers. In the case of a higher thickness than 1.6 mm, the soldering time is increased by 1 sec. for every commenced 0.8 mm of thickness beyond 1.6 mm.

Beispiel:
Für eine 6-Lagenschaltung mit 1,8 mm Dicke wird eine Lötzeit von

$$3 + 1 + 1 = 5 \text{ sec.}$$

benötigt.

Example:
6 layers, 1.8 mm cause a soldering time of

$$3 + 1 + 1 = 5 \text{ sec.}$$

c. Die Beurteilung wird nach IPC-S-804, Punkt 3.7, vorgenommen.

c. Assessment is performed according to IPC-S-804, Item 3.7.

d. Es darf keine Netzbildung auftreten.

d. No webbing shall be found.

13.2 Beanspruchung beim Löten

a. Die Leiterplatte muß beim Maschinenlöten folgender Beanspruchung standhalten:

   Löttemperatur: 260 ± 5 °C
   Lötzeit: 10 ± 1 s.

b. Beim Löten dürfen sich weder Risse an der Bohrungskante noch in der Hülse bilden (Corner Cracking und Barrel Cracking), noch eine Delaminierung oder Blasenbildung im Laminat auftreten. Es darf sich kein Teil des Leiterbildes vom Substrat abheben. Vgl. Punkt 5.18 bis 5.20.

c. Auf der Lötstoppmaske, falls vorhanden, darf keine Netzbildung (Webbing) vorkommen. Vgl. Punkt 8.3 g und 9.3 g.

d. Durch Fehler in der Leiterplatte verursachte Blowholes (Ausbläser) für maximal 2 % der Lötverbindungen werden akzeptiert, wenn die Leiterplatte vor dem Löten 1 bis 2 Stunden bei 110 bis 120 °C getempert wurde.

13.2 Effect of Soldering

a. The board shall stand the following soldering conditions when mass soldered:

   Soldering temperature: 260 ± 5 °C
   Soldering time: 10 ± 1 sec.

b. Soldering shall not cause cracks in plated-through holes (corner cracking and barrel cracking), delamination or blistering in the laminate. No part of the pattern shall be lifted. Cf. Items 5.18 to 5.20.

c. No webbing shall be found on the solder mask, if any. Cf. Items 8.3 g and 9.3 g.

d. Blowholes caused by defects in the board can be accepted in max. 2 % of the solder joints. An immediately preceding stoving of the board for 1 to 2 hours at 110 to 120 °C is implied.

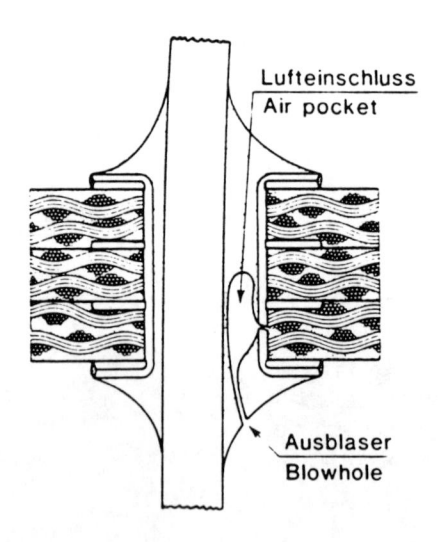

Lufteinschluss / Air pocket

Ausbläser / Blowhole

13.3 Solderability after Storage

a. Tin/lead plating with reflowing: min. 12 months
b. Solder coating and hot-air levelling: min. 12 months
c. Protective lacquering: min. 3 months
d. Immersion tin: min. 1 week

The above storage durabilities presume:
Storage Conditions

Temperature: 15 - 35 °C
Rel. humidity: 45 - 75 %
Storage*: in closed polyethylene bags

* Delivery packing is not included in the above specification and shall be agreed upon.

Solderability Test
The test is carried out as per Items 13.1 and 13.2.

14. Machining

14.1 General Requirements

a. The machining shall not cause crazing (cf. Item 1.5.3), delamination (cf. Item 1.5.5), haloing (cf. Item 1.5.6) or scratching of the board.

b. At punched holes, notches and slots a 0.3 mm haloing is accepted.

c. Along the edges of the board a slight delamination is acceptable provided it does not extend into the laminate by more than 0.3 mm from the edges.

14.2 Warp and Twist

14.2.1 Definitions (Rectangular Boards)

Warp is defined as the board's deviation from flatness. The warp has a cylindrical or spherical curvature, whilst all four corners of the board are in the same plane.

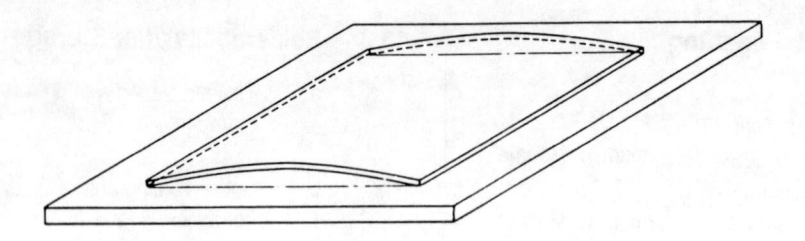

Die Wölbung kann in Längs- wie in Querrichtung einer Leiterplatte gleichzeitig auftreten, während die vier Leiterplattenecken aber in der gleichen Ebene bleiben.

Warp can occur lengthwise and crosswise at the same time whilst all four corners of the board are still in the same plane.

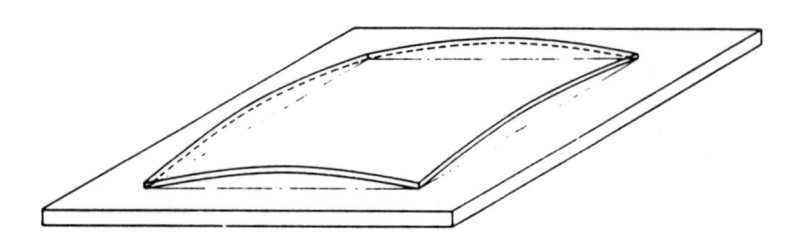

Die Verwindung (Twist) wird als Verformung einer Leiterplatte längs einer Diagonalen oder parallel zu dieser definiert. Dabei liegt eine Leiterplattenecke nicht in der gleichen Ebene wie die übrigen drei Ecken.

Twist is defined as a deformation of the board along a diagonal, or parallel to the diagonal, in such a way that one of the corners is not in the plane containing the other three corners.

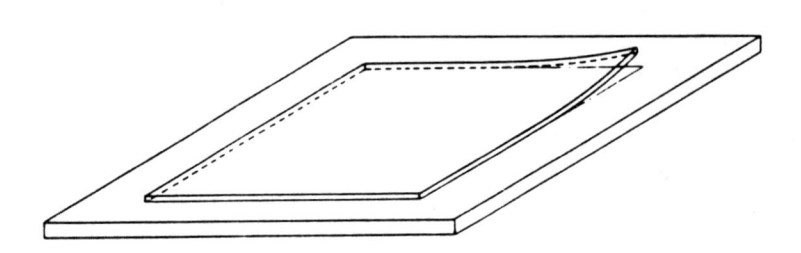

14.2.2 Bestimmungen von Wölbung und Verwindung

Die Wölbung wird als Prozentsatz K der Durchbiegung t_K bezogen auf die Länge L der durchgebogenen Kante bestimmt:

$$K = \frac{t_K}{L} \cdot 100\%$$

Wenn die Durchbiegung sowohl in Längs- als auch in Querrichtung auftritt, dann gilt der größere Wert.

14.2.2 Determination of Warp and Twist

Warp is determined as the percentage K of the deflection t_K relative to the length L of the curved edge:

$$K = \frac{t_K}{L} \cdot 100\%$$

If the warp occurs both lengthwise and crosswise, the largest value of warp found is used.

| Mechanische Bearbeitung | Machining |

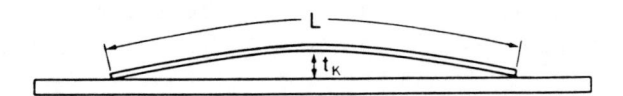

[Diagram: side view of a bowed board showing length L and deflection t_K]

| Verwindung (Twist) wird als Prozentsatz V der Verbiegung tv im Verhältnis zur Diagonalenlänge D angegeben: | Twist is determined as the percentage V of the deflection tv relative to the length D of the diagonal: |

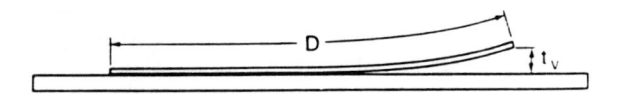

$$V = \frac{t_V}{D} \cdot 100\,\%$$

[Diagram: board with diagonal D and vertical deflection t_V at one corner]

| Wenn der Diagonalschnitt der Leiterplatte so, wie auf der nachstehenden Skizze erkennbar, aussieht, dann kann es erforderlich sein, zum Messen drei Leiterplattenecken aufzulegen. Die Verwindung (Twist) wird dann als Prozentsatz V der vertikalen Durchbiegung t_p im Verhältnis zur Diagonalenlänge D angegeben: | If the diagonal section of the board has an appearance as shown below, it may be necessary to support the three corners. The twist is then determined as the percentage V of the vertical displacement t_p relative to the length D of the diagonal: |

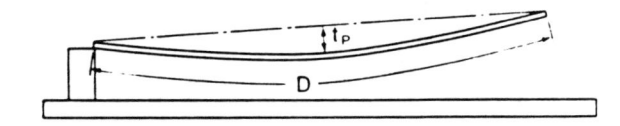

$$V = \frac{t_p}{D} \cdot 100\,\%$$

[Diagram: board supported at three corners showing diagonal D and vertical displacement t_p]

Die Messung von Wölbung und Verwindung wird in PERFAG 11 behandelt.	Inspection of warp and twist is addressed in PERFAG 11.
Hinweis	**Note**
Die Werte für Wölbung und Verwindung hängen davon ab, auf welcher Seite die Leiterplatte beim Messen liegt. Es ist in der Praxis jedoch üblich, die Messung so durchzuführen, wie es aus den Skizzen hervorgeht.	The value of the warp or twist depends upon which side of the board is turned upwards. However, when measuring the warp or twist, it is common practice to place the board as shown above

14.2.3 Normal zulässige Werte für Wölbung und Verwindung / Normal Warp and Twist Requirements

| Für die manuelle Leiterplattenbestückung und die in Absatz 1.1 genannten Laminattypen müssen für Leiterplatten mit einer Dicke t vor dem Löten folgende Forderungen eingehalten werden: | In the case of boards for manual assembly, made of the laminate types stated in Item 1.1, the warp and twist (before soldering) of finished boards with a thickness t must fulfil the requirements below: |

Machining

$0.8 \leq t < 1.5$ mm: $\leq 1.5\%$
$1.5 \leq t \leq 3.2$ mm: $\leq 1.0\%$

Note
Provided the PCB specification does not state a tighter requirement as to warp and twist, cf. Item 14.2.4, Note 1, the above requirements are valid.

14.2.4 Tighter Warp and Twist Requirements

To ensure a flawless automatic component insertion, tighter requirements are set to give a lower value of warp and twist (before soldering) of finished boards. Close collaboration between the PCB manufacturer and the PCB designer is therefore implied. In order to achieve the lowest possible warp and twist, the pattern should be balanced to achieve approximately the same pattern density on each side, and between the two sides.

The PCB manufacturer must carefully evaluate the PCB documentation received and make recommendations to balance the pattern to reduce the warp and/or twist to a minimum. Furthermore, he must adjust the manufacturing methods to minimize the warp and/or twist. For example, it can be necessary to use a special base material, to perform a heat stabilization or to choose a special orientation of the boards in the production panel.

In the case of boards made of the laminate types stated in Item 1.1, the goal is that the warp and twist (before soldering) of finished boards with a thickness t fulfil the requirements below. These requirements however, cannot be expected to be fulfilled by all PCB manufacturers.

$0.8 \leq t < 1.5$ mm: $\leq 1.0\%$
$1.5 \leq t \leq 3.2$ mm: $\leq 0.5\%$

Instead, some PCB manufacturers state the warp and twist according to a normal distribution curve with a standard deviation $s = 0.2\%$. For an average warp and twist of 0.5 % and a standard deviation of $s = 0.2\%$, the distribution of warp and twist within a batch will be:

Number of Boards	Warp and Twist
50 %	$\leq 0.5\%$
84 %	$\leq 0.7\%$
99 %	$\leq 1.0\%$

It is recommended that an agreement be reached with the PCB manufacturer regarding the actual specification.

Note 1
The PCB specification should state the tighter requirements of warp and twist only if necessary. If not, the requirements stated in Item 14.2.3 are followed.

Note 2
In the case of partly contoured boards in a panel format, the panel will normally exhibit a low value only of warp and twist, since the internal stress has been released because of the routing.

14.3 Reference System

The datum of the reference system is located in the centre of the primary reference hole R_1, and the direction of the X-axis is determined by the centre of the secondary reference hole R_2. The Y-axis is perpendicular to the X-axis. See figure in Item 14.4.

Note 1
Reference holes can be used as tooling holes or mounting holes for mechanical assembly. It is implied that reference holes, specified by the customer, are drilled or punched at the same time as the other holes of the board, and that their diameter is larger than 2 mm.

Design Note 1
Unless otherwise stated, reference holes are to be manufactured as nonplated-through holes. In this case, to ensure that tenting is successful, a clearance of min. 0.3 mm, preferably 0.5 mm, around the reference holes is required. Tenting of holes larger than 3.5 mm should not be specified without consulting the PCB manufacturer. Any scratches in the internal plating of plated-through reference holes, if any, caused by guide pins, are accepted.

Design Note 2
To help the PCB manufacturer inspect the reference holes (nonplated-through) and the circuit pattern registration, the design may include a metal ring around the reference holes. The width a of the metal ring may be 0.25 mm, and the clearance b may be 0.3 mm, preferably 0.5 mm.

Mechanische Bearbeitung | Machining

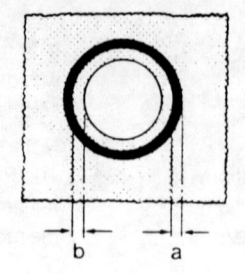

Designhinweis 3
Es ist nicht zweckmäßig, Referenzmarken innerhalb des Leiterbildes für das Referenzsystem zu verwenden. Solche Marken sollten nur dazu dienen, einen Kohledruck zum Leiterbild auszurichten. Vgl. 10.13.

Design Note 3
It is not expedient to base the reference system on reference marks in the pattern. Such marks should only be used in conjunction with carbon printing as an aid in achieving a correct registration of the carbon printing relative to the pattern. Cf. Item 10.13.

14.4 Bemaßung des Leiterplattenumrisses

Die Bemaßung des Leiterplattenumrisses einschließlich innenliegender Ausschnitte und von Schlitzen geht von der primären Referenzbohrung R_1 aus und gibt die Nennmaße an. Es sollten mindestens folgende Maße angegeben werden:

1. Die Bemaßung der Kontur durch die Einzelmaße X_1, X_2 ... und Y_1, Y_2 ... von der Referenzbohrung R_1 aus zu den Leiterplattenrändern.

14.4 Dimensioning of the Board Contour

Dimensioning of the board contour, including internal cut-outs and slots, is performed on the basis of the primary reference hole R_1 and states the nominal dimensions. The dimensioning should at least comprise:

1. A contour indication by the dimensions X_1, X_2, ... and Y_1, Y_2, ... from the reference hole R_1 to the board edges.

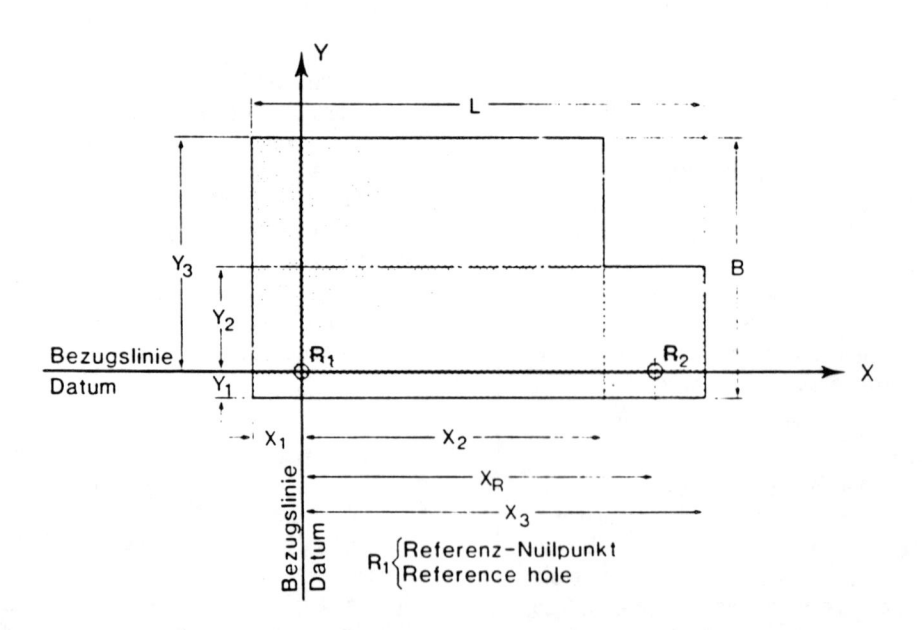

Hinweis
Der Nullpunkt des Referenzsystems kann in bestimmten Fällen außerhalb der Leiterplattenkontur festgelegt werden. In diesem Fall sollte das in der Leiterplattenspezifikation angegeben werden.

Note
The datum of the reference system can in certain cases be placed outside the board contour. If so, this should appear on the PCB specification.

Mechanische Bearbeitung	Machining
2. Alternativ kann der Umriß einer Leiterplatte auch durch die von der Referenzbohrung aus gemessenen Koordinaten (X_1, Y_1), (X_2, Y_2) ... der Eckpunkte einer Leiterplatte bestimmt werden.	2. Alternatively, a contour indication by the coordinates (X_1, Y_1), (X_2, Y_2) ... from the reference hole R1 to the board corners.

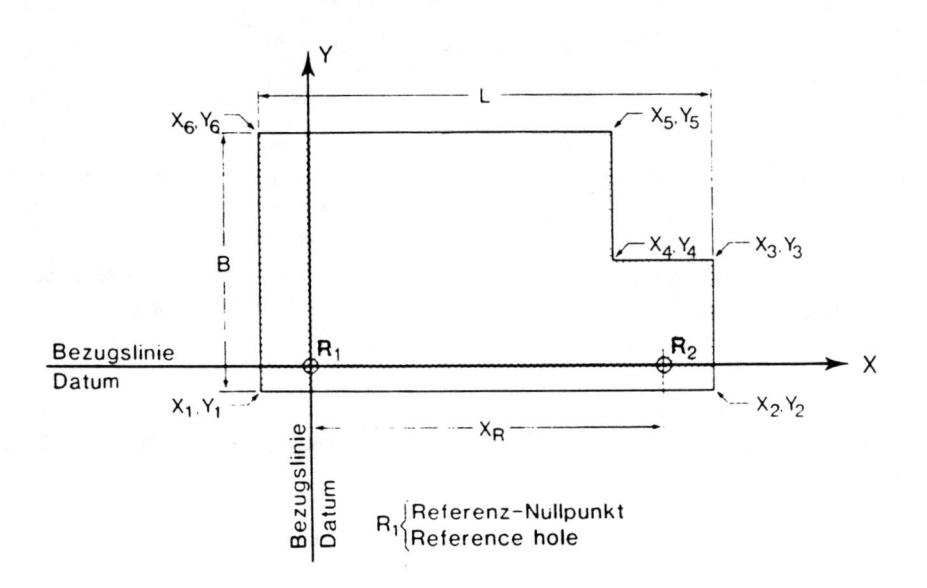

R_1 { Referenz-Nullpunkt / Reference hole }

3. Die Lagetoleranz des Umrisses sollte unter Berücksichtigung der mechanischen Bearbeitungstoleranz (s. Punkt 14.7) als generelle Toleranz angegeben werden.	3. The positional tolerance of the contour in accordance with the machining tolerances in Item 14.7, should preferably be stated as a general tolerance.
4. Bei innenliegenden Winkeln sollte ein maximaler Radius angegeben werden, vorausgesetzt, daß keine scharfe Ecke erforderlich ist. Ein praktikabler Radius ist $R \leq 1,2$ mm, d.h. daß ein 2,4 mm-Fräser verwendet werden kann.	4. In the case of inward corners, a max. radius of curvature should be stated, provided sharp corners are not required. An expedient indication is $R \leq 1.2$ mm, allowing the use of a 2.4 mm router bit.

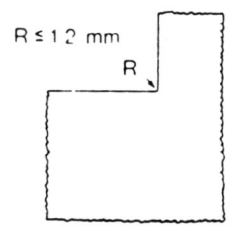

Designhinweis	Design Note
Falls scharfe Ecken benötigt werden, lassen sie sich wie unten skizziert herstellen.	Sharp corners, if required, can be obtained by routing as shown below.

| Mechanische Bearbeitung | Machining |

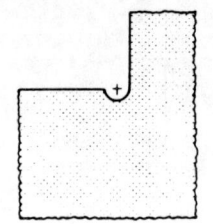

5. Zur Information können die nominale Gesamtlänge L und die Gesamtbreite B der Leiterplatte angegeben werden.
6. Der Nominalabstand X_R zwischen den beiden Referenzbohrungen R_1 und R_2.

5. For information, the overall dimensions of the board can be stated as its nominal length L and width B.
6. The nominal distance X_R between the two reference holes R_1 and R_2.

14.5 Bemaßung von im Nutzen angefertigen Leiterplatten

Gleiche Leiterplatten werden zu einem Nutzen addiert, um sie in einem größeren Plattenformat, d.h. kostengünstiger, fertigen zu können. Es ist möglich, diese Leiterplatten im Nutzen nur vorzutrennen (fräsen), um sie in größerer Anzahl zu bestücken und nachher endgültig zu trennen.

Hinweis
Beim Wellenlöten kann das Lot durch die gefrästen Schlitze dringen. Daher bevorzugen es einige Leiterplattenkunden, die Leiterplatten erst nach dem Bestücken, z.B. durch Wasserstrahlschneiden oder Stanzen, zu trennen. Es darf daher nur umrißgefräst werden, wenn dies in der Leiterplattenspezifikation vorgeschrieben ist.

14.5 Dimensioning of Panelized Boards

Panelization means the placing of several similar boards in a production panel, which can be delivered as a contoured panel in which the individual boards are contoured except for some remaining tabs holding the boards in the panel. In this way it is possible to assemble a large number of boards in the same operation.

Note
In the case of wave soldering, the solder can rise through the slots. Therefore, some PCB customers may want to separate the boards after assembly, e.g., by waterjet cutting or punching. Contouring shall be performed only when specifically stated in the PCB specification.

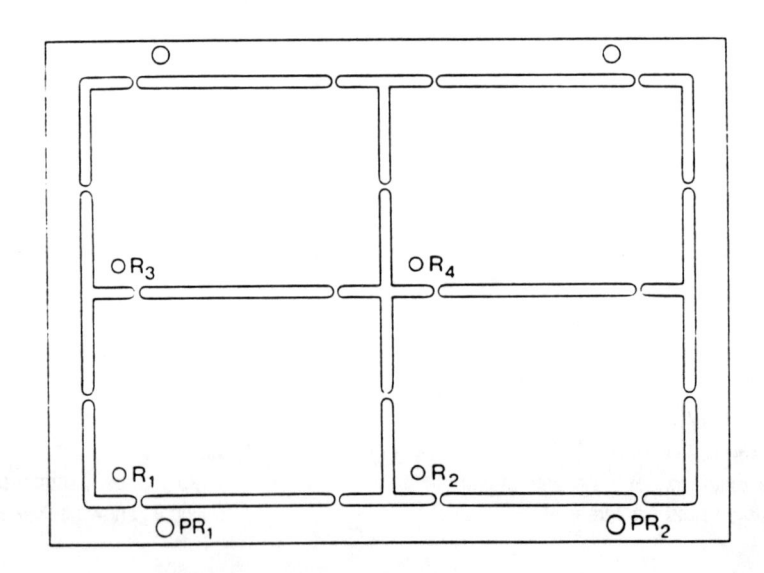

Designhinweis 1	Design Note 1
Beim Trennen der Leiterplatten können Basismaterialreste stehenbleiben. Dies kann dadurch verhindert werden, daß die Fräserbahn gemäß untenstehender Skizze angelegt wird.	Separation of the boards leaves remnants of the tabs protruding. This can be avoided by carrying out the router path as shown below.

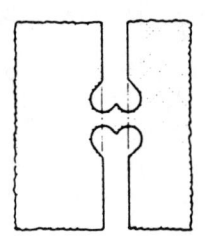

In gleicher Art und Weise wie in Abschnitt 14.3 erwähnt, können primäre und sekundäre Referenzbohrungen PR_1 und PR_2 auch auf einem Nutzen angebracht werden. Die einzelnen Leiterplatten im Nutzen werden mit Bezug auf die primäre Nutzenreferenzbohrung PR_1 bemaßt. Es reicht dann aus, die Lage der einzelnen Leiterplatten im Nutzen durch ihre primären Referenzbohrungen R_1, R_2, R_3 ... anzugeben, so daß der Leiterplattenhersteller selbst die Nutzenaddition durchführen kann.

Der nominale Abstand zwischen den einzelnen Leiterplatten beträgt üblicherweise 2,4 oder 3,0 mm. Im Normalfall wird mit einem 2,4 mm-Fräser gefräst, so daß im ersten Fall nur 1 Durchgang erforderlich ist. Dabei tritt auf einer Seite des Fräsweges eine leichte Krümmung der Leiterplattenkante auf. Durch einen zweiten Fräsdurchgang läßt sich eine höhere Genauigkeit erreichen.

Haltestege sind üblicherweise 1 mm breit. Bei gestanzten Leiterplatten soll ihre Breite wenigstens der Leiterplattendicke entsprechen, mindestens aber 1 mm betragen.

In the same way as mentioned in Item 14.3, primary and secondary panel reference holes PR_1 and PR_2 can be introduced on the panel. The individual boards are dimensioned relative to the primary panel reference hole PR_1. It is sufficient to indicate the location of the individual boards in the panel by their primary reference holes R_1, R_2, R_3 ...., thereby allowing the PCB manufacturer to take care of the panelization.

The nominal distance between the individual boards will usually be 2.4 mm or 3.0 mm. Normally a standard 2.4 mm router bit will be used, for which reason the routing in the first case is performed only once, resulting in a slightly larger unevenness at one side of the path. In the second case the individual board is routed in two passes, which gives a slightly greater evenness of the edges and a better accuracy.

A common width of the fixing tabs is 1 mm. In the case of blanked boards, the fixing tabs should have a width of no less than the thickness of the board, minimum 1 mm.

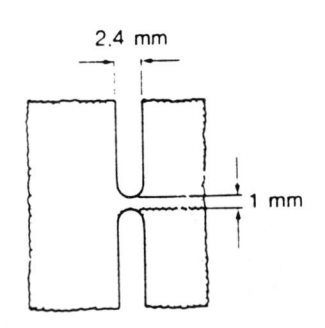

Mechanische Bearbeitung	Machining

Designhinweis 2
Leiterplattenhersteller verwenden unterschiedliche Nutzenformate. Daher sollte sich der Leiterplattendesigner mit dem Hersteller zwecks optimaler Ausnutzung des Nutzenformates in Verbindung setzen.

Designhinweis 3
Für die automatische Bestückung im Nutzen sollten Freiflächen am Nutzenrand sowie Aufnahmebohrungen vorgesehen werden, wie unter 4.12 a und b angegeben. Als Aufnahmebohrungen können auch die Nutzenreferenzbohrungen PR_1 und PR_2 verwendet werden.

Designhinweis 4
Wenn nach dem Löten ein maschinelles Abschneiden der Drahtenden vorgesehen ist, so ist ein vorgetrennter Nutzen hierfür im Normalfall nicht starr genug. Daher wird empfohlen, breitere Stege (ca. 20 mm) zur Erhöhung der Stabilität vorzusehen.

14.6 Festlegung des Umrisses

Der Leiterplattenumriß kann auf 3 Arten festgelegt werden. Aus der Leiterplattenspezifikation muß die Bemaßungsmethode ebenso hervorgehen wie der Hinweis, ob der Kunde Fräsdaten zur Verfügung stellt.

a. Die Bemaßung kann von einem Referenzsystem ausgehen, vgl. 14.3, und nach den in 14.4, Unterpunkt 1, genannten Bemaßungsregeln alternativ auch über die Koordinaten der Leiterplattenecken erfolgen, s. 14.4, Unterpunkt 2.

b. Fräsdaten des Kunden mit einem PERFAG 10 entsprechenden Format.

   Bemerkung
   Wenn Differenzen zwischen den Fräsdaten und der Bemaßung (vgl. 14.4 und 14.5) auftreten, gelten immer die Fräsdaten. Der Leiterplattenhersteller muß den Kunden über Differenzen informieren.

c. Eckenmarkierungen auf den Filmen, vgl. 4.1 a und b. In diesem Fall muß sich der Leiterplattenhersteller selbst ein Referenzsystem schaffen, vgl. 14.3, und, basierend auf diesem Referenzsystem, den Leiterplattenumriß digitalisieren. Der Kunde ist berechtigt, diese Daten zur Wareneingangskontrolle zu benutzen.

Design Note 2
Since many PCB manufacturers use production panels of various sizes, the PCB designer should consult the PCB manufacturer regarding details of the panelization in order to utilize the production panel in the best possible way.

Design Note 3
For the sake of automatic component assembly, the panelized boards should have edge clearance areas and tooling holes as mentioned in Items 4.12 a and b. The tooling holes can be the same as the panel reference holes PR_1 and PR_2.

Design Note 4
If automatic lead cutting is to be performed after soldering the board, the rigidity of the panelized boards will usually be insufficient to make this possible. Therefore, it is recommended to incorporate bars, e.g. with a width of 20 mm, in the panel in order to achieve a higher rigidity of the panel.

14.6 Determination of Contour

The contour of the board can be determined in three ways. The PCB specification shall indicate which method to follow, and it shall also indicate whether the customer's router data is enclosed.

a. Dimensioning can be based on a reference system, cf. Item 14.3 and according to the rules of dimensioning, cf. Item 14.4, Subitem 1, alternatively by coordinates of the board's corners, cf. Item 14.4, Subitem 2.

b. The customer's router data with a format according to PERFAG 10.

   Note
   In the case of known differences between the router data and the dimensioning, cf. Items 14.4 and 14.5, the router data always prevails. The PCB manufacturer shall inform the customer about the discrepancies.

c. Corner marks on the films, cf. Item 4.1, Subitems a and b. In this case, the PCB manufacturer himself shall establish a reference system, cf. Item 14.3, and based on the reference system, digitize the board's contour. The customer is entitled to be supplied with this data for use when inspecting the delivered boards.

Mechanische Bearbeitung / Machining

14.7 Bearbeitungstoleranzen

Für die Umrißposition gelten (in Bezug auf die nominale Position) folgende Toleranzen sowohl für einzelne als auch für im Nutzen addierte Leiterplatten bis zu einer Größe von 450 x 450 mm.

Fräsen und Stanzen	± 0,1 mm
Sägen und Schneiden	± 0,3 mm
Ritzen (Kerben)*	± 0,4 mm

* nach dem Trennen der Leiterplatten

Der Umriß wird anhand des Referenzsystems kontrolliert, vgl. Abschnitt 14.3.

Hinweis

Wenn keine Toleranz für die Lage des Umrisses angegeben ist, vgl. 14.4, Unterpunkt 3, kann der Leiterplattenhersteller selbst die zweckmäßigste Bearbeitungsmethode wählen.

14.8 Umrißbearbeitung

Zusätzlich zur Außenkontur sind auch innenliegende Ausschnitte und größere Bohrungen, die weder gestanzt noch gebohrt werden können, herzustellen.

Designhinweis

Der nominale Abstand zwischen der Leiterplattenaußenkante und Metallflächen, Leitern, usw. soll, wenn der Umriß gefräst wird, normalerweise mindestens 0,5 mm betragen. Das gilt auch für innenliegende Ausschnitte, Schlitze und größere Bohrungen.

14.9 Ritzen

Wenn Leiterplatten im Nutzen geliefert werden, in dem die einzelnen Leiterplatten geritzt sind, kann auch, falls nichts anderes vereinbart, die Außenkontur des Nutzens geritzt sein.

Der Abstand zwischen der nominalen Mittellinie der Kerbe C und metallenen Bereichen, Leitern, usw. soll normalerweise nicht kleiner als 1,2 mm sein.

Für das Ritzen von 1,6 mm dicken Leiterplatten gelten folgende Werte:

Kerbwinkel V:	30 - 45°
Versatz F* der Ritzung:	± 0,25 mm
Versatz P* des Umrisses:	± 0,40 mm
(nach dem Trennen der Leiterplatten)	

14.7 Machining Tolerances

The following tolerances regarding the position of the contour (relative to the nominal position) are valid for individual boards, or boards delivered in panels, up to 450 x 450 mm in size.

Routing and punching:	± 0.1 mm
Sawing and shearing:	± 0.3 mm
Scoring*:	± 0.4 mm

* after board separation

Inspection of the contour is based on the reference system, cf. Item 14.3.

Note

When no tolerance is given for the contour position, cf. Item 14.4, Subitem 3, the PCB manufacturer is given a free hand to choose the best method of machining the contour.

14.8 Machining the Contour

In addition to the outer contour, the machining also comprises internal cut-outs and larger holes which cannot be drilled or punched.

Design Note

The nominal distance between the board's contour and metal areas, conductors, etc., shall not normally be less than 0.5 mm when the contour is to be routed. This applies also to internal cut-outs, slots and larger holes.

14.9 Scoring

In the case of boards delivered in panels where the individual boards are scored, the outer contour of the panels can also be scored, unless stated otherwise in the PCB specification.

The distance between the nominal centre line C of the score and metal areas, conductors, etc., should normally not be less than 1.2 mm.

The specifications below are valid when scoring a 1.6 mm board:

Angle V of score:	30 - 45°
Displacement F* of score:	± 0.25 mm
Displacement P* of contour:	± 0.40 mm
(after board separation)	

Ritztiefe S:	mind. 0,10 mm	Scoring depth S:	min. 0.10 mm
Kerndicke K:	0,45 ± 0,10 mm	Core thickness K:	0.45 ± 0.10 mm
* In Bezug auf die nominale Mittellinie C		* Relative to the nominal centre line C	

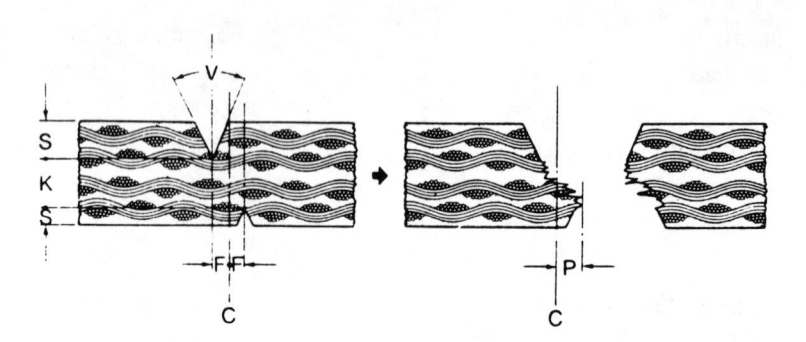

Die Lage der Ritzkontur wird anhand des Referenzsystems kontrolliert, vgl. 14.3.

Designhinweis

FR4-Laminate können zwar geritzt werden, aber die Glasfasern bereiten beim Trennen der Leiterplatten Schwierigkeiten, da die Kanten rauh werden und Glaspartikel abbrechen. Deshalb wird das Ritzen von FR4-Material nicht empfohlen.

Inspection of the position of the scoring is based on the reference system, cf. Item 14.3.

Design Note

FR4 laminates can be scored but the glass fibres cause problems in board separation, uneven edges and environmental problems. Scoring of FR4 materials is therefore not recommended.

14.10 Kodierschlitze in Randkontaktleisten

14.10.1 Lage des Schlitzes

Die Lage des Schlitzes kann festgelegt werden durch:

a. Das Nennmaß X_1 im generellen Referenzsystem X_G, Y_G, vgl. 14.3.

b. Das Nennmaß X_2 in einem lokalen Referenzsystem X_L, Y_L, das parallel zum generellen Referenzsystem X_G, Y_G liegt und seinen Nullpunkt in R_3 hat, vgl. 14.3.

c. Symmetrische Lage zwischen benachbarten Kontaktfingern, d.h. $X_3 = X_4$.

14.10 Edge Connector Polarization Slot

14.10.1 Position of the Slot

The position of the slot can be determined by:

a. The nominal dimension X_1 in the general reference system X_G, Y_G, cf. Item 14.3.

b. The nominal dimension X_2 in a local reference system X_L, Y_L parallel with respect to the general reference system X_G, Y_G, cf. Item 14.3, and with its datum in the local reference hole R_3.

c. Demand for a symmetrical location with respect to adjacent contact tabs, i.e., $X_3 = X_4$

Mechanische Bearbeitung / Machining

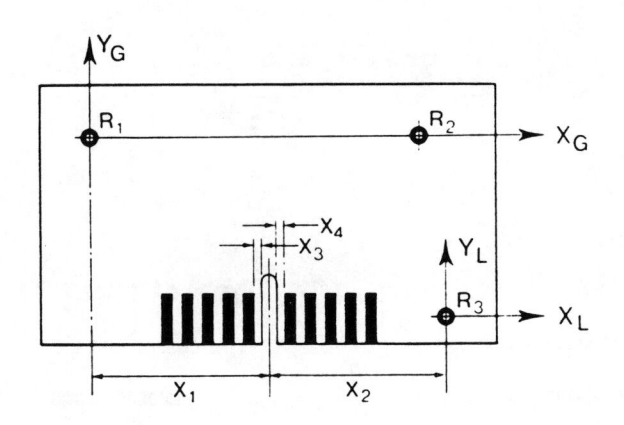

14.10.2 Herstellung des Kodierschlitzes

Fräsen oder Sägen

Breitentoleranz	± 0,10 mm
Längentoleranz	± 0,25 mm
Versatz der Schlitz-Mittellinie	≤ 0,10 mm

Hinweis

Wegen der Bearbeitungstoleranzen sollte der Abstand vom Schlitzrand zum Nachbarkontakt mindestens 0,3 mm betragen.

14.10.2 Machining of the Slot

Milling or sawing

Width tolerance:	± 0.10 mm
Length tolerance:	± 0.25 mm
Displacement of centre line of slot:	≤ 0.10 mm

Note

Because of manufacturing tolerances, the nominal distance from the slot edge to the adjacent contact tab should not be less than 0.3 mm.

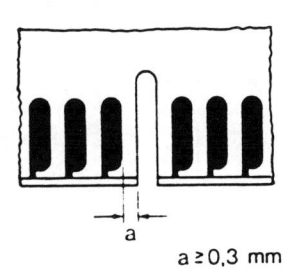

$a \geq 0.3$ mm

14.11 Anfasung der Randkontakte

Randkontakte sollen mit einer 45°-Fase versehen werden. Für die Maße a und b gelten pro Kante folgende Werte:

$0{,}3 \leq a \leq 0{,}4$ mm, für $t \geq 1{,}6$ mm
$0{,}3 \leq b \leq 0{,}4$ mm, für $t \geq 1{,}6$ mm

Bei der Bearbeitung darf sich weder das Kupfer lösen noch darf ein sichtbarer Grat entstehen.

14.11 Chamfering of Edge Connector

The edge connector contacts shall be chamfered at a nominal angle of 45°. The chamferings a and b per edge shall meet the following requirements:

$0.3 \leq a \leq 0.4$ mm, when $t \geq 1.6$ mm
$0.3 \leq b \leq 0.4$ mm, when $t \geq 1.6$ mm

No visible burrs or lifting of the copper at the machined surface of the contacts shall occur.

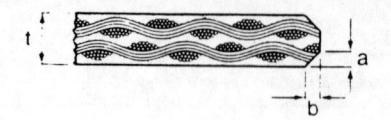

14.12 Locharten

Folgende Locharten können unterschieden werden:

a. Bohrungen für die manuelle Bestückung (HMT)
b. Bohrungen für die automatische Bestückung (HMT)
c. Aufnahmebohrungen für die automatische Bestückung
d. Befestigungsbohrungen bzw. -löcher

14.12.1 Bohrungen für die manuelle Bestückung

Wenn in der Leiterplattenspezifikation nicht anders angegeben, werden die Löcher in der Leiterplatte zum manuellen Bestücken von Lochmontagebauelementen verwendet.

14.12.2 Bohrungen für die automatische Bestückung

Die automatische Bestückung stellt hohe Anforderungen bezüglich der Lagegenauigkeit der Bohrungen in einer Leiterplatte. Das setzt eine enge Zusammenarbeit zwischen dem Leiterplattendesigner und dem Leiterplattenhersteller voraus.

Der Leiterplattenhersteller muß die Qualität der angelieferten Fertigungsunterlagen sorgfältig prüfen. Darüberhinaus muß er die Fertigungsprozesse so steuern, daß prozeßbedingte Abweichungen minimiert werden. Zum Beispiel kann eine Wärmebehandlung des Basismaterials zur Dimensionsstabilisierung erforderlich sein.

14.12.3 Aufnahmebohrungen für die automatische Bestückung

Die Anforderungen an Aufnahmebohrungen für die automatische Bestückung sind dieselben, die auch für die Bohrungen zum automatischen Bestücken von Lochmontage-Bauelementen gelten, vgl. Punkt 14.12.2.

Für die Aufnahmebohrungen wird die Tentingtechnik in der gleichen Art angewandt, wie sie unter Punkt 14.3 für die Referenzbohrungen beschrieben wird (vgl. Punkt 14.3, Bemerkung 1 und Designhinweis 1).

14.12 Types of Holes

In the case of holes, distinction should be made

a. Holes for manual assembly (HMT)
b. Holes for automatic assembly (HMT)
c. Tooling holes for automatic assembly
d. Mounting holes

14.12.1 Holes for Manual Assembly

Unless otherwise stated in the PCB specification, the holes of the board are used for manual assembly of HMT components.

14.12.2 Holes for Automatic Assembly

To ensure a flawless component insertion, there are strict requirements concerning the positional accuracy of the holes in the finished board. Close collaboration between the PCB manufacturer and the PCB designer is therefore implied.

The PCB manufacturer must carefully evaluate the quality of the PCB documentation received. Furthermore, he must adapt the manufacture to minimize process deviations. For example, it may be necessary to stove the base material prior to manufacture in order to reduce the permanent dimensional change.

14.12.3 Tooling Holes for Automatic Assembly

The demands on tooling holes for automatic component assembly are the same as for holes for automatic assembly of HMT components, cf. Item 14.12.2.

The tooling holes are produced using a tenting technique in the same way as described for the reference holes of the board's reference system, cf. Item 14.3, Note 1 and Design Note 1.

Mechanische Bearbeitung	Machining

14.12.4 Montagebohrungen

Montagebohrungen werden zur Befestigung der Leiterplatte oder Befestigung bestimmter Bauelemente auf der Leiterplatte benutzt.

Metallisierte Montagebohrungen werden gemeinsam mit den HMT-Bohrungen gebohrt. Für nicht metallisierte Montagebohrungen wird die Tentingtechnik in der gleichen Art angewandt, wie sie unter Punkt 14.3 für die Referenzbohrungen beschrieben wird (vgl. Punkt 14.3, Bemerkung 1 und Designhinweis 1).

Nicht metallisierte Montagebohrungen können jedoch auch in einem späteren Arbeitsgang gebohrt werden.

14.13 Festlegung der Lochpositionen

Die Lochposition kann festgelegt werden durch:

a. Die Bohrdaten des Kunden, vorausgesetzt, daß diese aus derselben Datenbank wie die Plotterdaten generiert wurden. Das Datenformat kann entsprechend PERFAG 10 gewählt oder vereinbart werden.

Hinweis
Wenn Differenzen zwischen dem Bohrprogramm und den Originalfilmen auftreten, vgl. 4.1, muß der Leiterplattenhersteller mit dem Kunden Kontakt aufnehmen.

b. Die Koordinatenangaben des Kunden, basierend auf dem Referenzsystem der Leiterplatte, vgl. 14.3. Der Nullpunkt des Referenzsystems kann in bestimmten Fällen außerhalb der Leiterplattenkontur liegen. Dies soll aus den Fertigungsunterlagen hervorgehen.

c. Bohrungsmarkierungen auf den Original-Kundenfilmen entweder als Lötaugen oder als Markierung der Bohrungszentren. In beiden Fällen muß der Leiterplattenhersteller die Bohrdaten durch eine Digitalisierung der Filme gewinnen. Der Kunde ist berechtigt, diese Daten für Kontrollzwecke zu benutzen.

14.14 Lagetoleranzen für Bohrungen

Bei Leiterplatten für die Lochmontagetechnik besteht die Hauptanforderung darin, daß der relative Versatz zwischen Bohrung und Lötauge so klein ist, daß die Anforderungen bezüglich eines vollständig geschlossenen Restringes innerhalb der angegebenen Toleranzen erfüllt werden.

14.12.4 Mounting Holes

Mounting holes are used for the mechanical fixing of the board or the fixing of certain components on the board.

Plated-through mounting holes are drilled at the same time as the HMT holes. Nonplated-through mounting holes are produced using a tenting technique in the same way as described for the reference holes of the board's reference system, cf. Item 14.3. Note 1 and Design Note 1.

Nonplated-through mounting holes, however, can also be drilled in a later drilling operation.

14.13 Determination of Hole Position

The position of the holes can be determined by:

a. The customer's drill data, assuming it be based on the same data base as used when generating the plotter data. Format according to PERFAG 10 or to a previous arrangement

Note
In the case of known differences between the drill tape and the original films, cf. Item 4.1, the PCB manufacturer shall contact the customer.

b. The customer's coordinate indications based on the reference system of the board, cf. Item 14.3. The datum of the reference system can in certain cases be placed outside the board contour. If so, this should appear in the PCB specification.

c. Hole markings on the customer's original films, either as solder pads or as centre markings. In both cases, the PCB manufacturer shall determine the drill data by digitizing the films. The customer is entitled to be supplied with this data for use when inspecting the delivered boards

14.14 Positional Tolerances on Holes

The main requirement of HMT holes is that the relative displacement between hole and solder pad be so small that the requirements regarding an unbroken annular ring, including the stated exceptions, are met.

| Mechanische Bearbeitung | Machining |

Darüberhinaus gelten folgende Anforderungen:
Der radiale Bohrungsversatz ΔF in beliebiger Richtung muß für 200 x 200 mm große Leiterplatten kleiner als die unten angegebenen Werte sein. Bei größeren Leiterplatten werden 0,05 mm je begonnene 100 mm über 200 mm hinaus hinzuaddiert. Dies gilt für Leiterplattengrößen bis zu 450 x 450 mm.

Bohrungen für die manuelle Bestückung:	± 0,15 mm
Bohrungen für die automatische Bestückung:	± 0,10 mm
Aufnahmebohrungen, die mit den Bauelementebohrungen zusammen gebohrt oder gestanzt werden:	± 0,10 mm
Befestigungsbohrungen, die mit den Bauelementebohrungen zusammen gebohrt oder gestanzt werden:	± 0,15 mm
In einem späteren Bearbeitungsschritt gebohrte Montagebohrungen:	± 0,20 mm

Für Referenzbohrungen gilt der unten angegebene radiale Versatz ΔF für Leiterplattengrößen bis zu 450 x 450 mm:

Primäre Referenzbohrung R_1: (per Definition)	0,00 mm
Sekundäre Referenzbohrung R_2:	± 0,05 mm

Kontrollmessungen
Der Versatz ΔF wird auf der Basis des Referenzsystems, vgl. 14.3 und der Bohrdaten, vgl. 14.13, gemessen.

Designhinweis
In der Leiterplattenspezifikation sollte die Lagetoleranz, wie unten dargestellt, angegeben werden, wobei die zusätzliche Toleranz für Leiterplatten über eine Größe von 200 x 200 mm hinaus akzeptiert wird.

Für manuell zu bestückende Leiterplatten sollte angegeben werden:
> Lagetoleranz der Bohrungen:
> ± 0,15 mm innerhalb 200 x 200 mm Leiterplattenfläche

Für automatisch zu bestückende Leiterplatten sollte angegeben werden:
> Lagetoleranz der Bohrungen:
> ± 0,10 mm innerhalb 200 x 200 mm Leiterplattenfläche.

Furthermore, the following requirements are valid:
The radial displacement ΔF of a hole in any direction shall be less than stated in the survey below, valid for board sizes up to 200 x 200 mm. In case of larger boards, 0.05 mm is added for every additional 100 mm of length beyond 200 mm. This applies to board sizes up to 450 x 450 mm.

Holes for manual assembly:	± 0.15 mm
Holes for automatic assembly:	± 0.10 mm
Tooling holes drilled or punched together with the HMT holes:	± 0.10 mm
Mounting holes drilled or punched together with the HMT holes:	± 0.15 mm
Mounting holes drilled in a later drilling operation:	± 0.20 mm

For reference holes the radial displacements ΔF are valid for board sizes up to 450 x 450 mm:

Primary reference hole R_1: (per definition)	0.00 mm
Secondary reference hole R_2:	± 0.05 mm

Inspection
The displacement ΔF is determined on the basis of the reference system, cf. Item 14.3, and the drill information according to Item 14.13.

Design Note
The PCB specification should state the positional tolerance as shown below, the additional tolerance for boards beyond 200 x 200 mm being implied.

In the case of boards for manual assembly, the tolerance should be specified as:
> Positional tolerance on holes:
> ± 0.15 mm within 200 x 200 mm

In the case of boards for automatic assembly, the tolerance should be specified as:
> Positional tolerance on holes:
> ± 0.10 mm within 200 x 200 mm

Stichwortverzeichnis

A

Abdeckung
- des Leiterbildes, mit abziehbaren Masken 69
- von Leitern 56
  * mit Fotopolymermasken 56
  * mit Siebdruckmasken 51
- von Lötaugen bzw. -flächen,
  mit abziehbaren Masken 69
- von Seitenkanten, von Randkontakten 45

Abdeckung der Lochkante,
  bei nichtmetallisierten Bohrungen 44

Abheben, von Kupfer 26, 43

Ablösen, der Galvanikschicht, von Randkontakten 45

Ablösung der Hülse von der Lochwand,
  in metallisierten Bohrungen 40

Abziehbare Masken
- Abdeckung
  * des Leiterbildes 69
  * von Lötflächen, auf der Gegenseite 69
- Anwendungszweck 67
- auf Kohleelemente 65
- Ausführung 67
- Entfernung, der Maske 68
- Erwärmung, Auswirkungen 68
- Fehlstellen, unerwünschte 70
- Haftfestigkeit 67
- Kantenschärfe 68
- Lösungsmittel, Beständigkeit gegen 68
- Qualitätsanforderungen 67
- Verschließen, von Bohrungen 69

Additivtechnik 9

Ätzfaktor, Anforderungen 22

Anfasen, von Randsteckkontakten 85

Anordnung der Lagen 10

Anordnung, des Komponentendrucks 50

Anschlußflächen, für den Kohledruck 61

AOI (Automatic Optical Inspection) 30

Arbeitsfilme, mit Kompensation 19

Aufbau, von fertigen Leiterplatten 10

Aufnahmebohrungen,
  für die automatische Bestückung 29

Aufquellen, lokales, des Basismaterials 12

Aufschmelzen,
- Abdecken
  * des Übergangs Bohrung - Lötauge 18
  * von Leiterkanten 17
- Ausführung des 18
- bei Entnetzung 17
- bei Nichtbenetzung 17
- Qualitätsanforderungen, an die Oberfläche 17
- Unvollständiges 17
- Verfärbung, des Basismaterials 17
- von Zinn/Blei 17

Ausbluten, von Siebdruckmasken 48

Ausführung, von abziehbaren Masken 67

Ausfüllen, von Zwischenräumen zwischen Leitern
- mit Fotopolymer-Flüssigfilmmasken 55
- mit Fotopolymer-Trockenfilmmasken 54
- mit Siebdruckmasken 51

Aushärten
- von Fotopolymermasken 54
- von Kohledruck 66
- von Siebdruckmasken 49

Ausrichtung der Ober- zur Unterseite,
  bei Randkontakten 47

Aussparungen, auf Innenebenen 32

Automatische Bestückung
- Aufnahmebohrungen 29
- Freiflächen am Rand 28
- Registriermarken (Fiducials) 29
- Registriermarken, optische 29
- SMT-Leiterplatten 28

Automatische optische Kontrolle 30

B

Basismaterial
- Aufquellen 12
- Aushärten 14
- Blasenbildung (Blistering) 12
- Delaminierung 12
- Hofbildung (Haloing) 13
- metallische Einschlüsse 15
- mit Gewebefreilegung (weave exposure) 14
- mit Gewebestrukturbildung (weave texture) 14
- Typen 8

Basismaterial, starr 10

Bearbeitungstoleranzen
- beim Abscheren 83
- beim Fräsen 83
- beim Ritzen 83
- beim Sägen 83
- beim Stanzen 83

Bemaßung
- des Leiterplattenumrisses 78
- von im Nutzen angeordneten Leiterplatten 80

Blasenbildung, im Basismaterial 12

Bohrungen, Arten 86

Stichwortverzeichnis

C

CAD-Plotterdaten	18
Chemische Verzinnung	71
Codierschlitze, in Randkontaktleisten	
- Anordnung	84
- Herstellung	85
- Lage	84

D

Delaminierung	
- entlang der Leiterplattenkanten	73
- im Basismaterial	43
- nach der mechanischen Bearbeitung	73
- unter Fotopolymermasken	54
- unter Siebdruckmasken	49
- von Basismaterial	13
• lokal	13
- von Kohleelementen	67
Detaillierungsgrad, des Kohleleiterbildes	57
Dichte, gleichmäßige, des Leiterbildes	19
Dicke	
- der fertigen Leiterplatte	8
• Messung	8
- von Fotopolymer-Flüssigfilmmasken	55
- von Fotopolymer-Trockenfilmmasken	54
- von Gold	17
- von Kupfer	
• auf Leitern	16
• in metallisierten Bohrungen	15
- von Kupferfolie	8
• für den Kohledruck	56
- von Siebdruckmasken	49
- von steckbaren Leiterplatten	8
Dickentoleranz, von fertig bearbeiteten Leiterplatten	8
Dielektrisches Material	
- Dicke	10
- Dickentoleranz	11
Dimensionsänderungen, des Leiterbildes, permanente	26
Dokumentation, Beurteilung der	21
Druckbreite, Kohledruck	58
Durchmessertoleranz	
- von metallisierten Bohrungen	31
- von nichtmetallisierten Bohrungen	44

E

Ecken, innenliegend, Kurvenradius	79
Einbuchtungen	
- entlang Leiterkanten	24
- von Kohledruck	59
Elektrische Prüfungen	30
Entfernung	
- von abziehbaren Masken	68
- von Fotopolymermasken	54
Entnetzung, nach dem Aufschmelzen	17
Epoxid-Verschmierung (smear), in metallisierten Bohrungen	36
Erstarrungslinien, in Zinn/Blei, nach dem Aufschmelzen	17
Erwärmung, Auswirkungen auf abziehbare Masken	68

F

Fehlstellen	
- im Kohledruck	
* auf dem Basismaterial	60
* auf Kupfer	60
- im Leiterbild	25
- in abziehbaren Masken	70
- in metallisierten Bohrungen	35
Filme	
- abgeleitet aus CAD-Plotterdaten	18
- fotografisch modifiziert	18
Fleckenbildung (Measling) im Basismaterial	11
Fotopolymermasken	
- Anwendung	52
- Aushärtung, der Maske	54
- Delaminierung, unter der Maske	54
- Entfernung, von Masken	54
- Ionische Verunreinigung, unter der Maske	53
- Kupfer, blank/oxidiert, unter der Maske	53
- Lösungsmittel, Beständigkeit gegen	53
- Lufteinschlüsse, unter der Maske	54
- Materialien	54
- Metallpartikel, auf oder unter der Maske	53
- Qualitätsanforderungen	53
- Risse, in der Maske	53
- Rückstände, der Maske	53
- Tenting, von Durchverbindungen	53
- Zinn/Blei, unter der Maske	53
Fotowerkzeuge, mit Kompensation	19
FR-4	8
Freiflächen am Rand, für die automatische Bestückung	28
Fünfundsiebzig-Prozent-Regel	
- beim Kohleleiterbild	58
- beim Kupferleiterbild	19

Stichwortverzeichnis

G

Galvanischer Niederschlag	
- bei Gold	17
* auf Nickel	16
- bei Kupfer	15
- Gleichmäßigkeit, des Leiterbildes	20
- Haftfestigkeit	15
- mit Zinn/Blei	16
* Legierungszusammensetzung	16
- Qualitätsanforderungen	15
- "Verbrennen"	15
Galvanisiermaterialien, für Randsteckkontakte	16
Gewebefreilegung, des Basismaterials	14
Gewebestrukturbildung, des Basismaterials	14
Gewebezerrüttung (Crazing)	
- im Basismaterial	12
- nach der mechanischen Bearbeitung	73
Glasfasern, in metallisierten Bohrungen	30
Gratbildung	
- bei nichtmetallisierten Bohrungen	44
- in metallisierten Bohrungen	40

H

Haftfestigkeit, von	
- abziehbaren Masken	67
- galvanischen Niederschlägen	
* des Leiterbildes	15
* von Randkontaktleisten	47
- Kohledruck	63
- Leitern	65
- SMD-Lötflächen	26
Haltestege, für im Nutzen angeordnete Leiterplatten	80
Haltestege, für Leiterplattennutzen	80
Harzrückzug, in metallisierten Bohrungen	40
Heißluftverzinnung (Hot-Air-Levelling)	70
Hofbildung (Haloing)	
- im Basismaterial	13
- nach dem Stanzen	73
- nach der mechanischen Bearbeitung	73

I

Im Nutzen angeordnete Leiterplatten	
- Abstände	80
- Bemaßung	80
- Haltestege	80
Innenlagen	
- Ebenen	
* Abstand, zwischen Lötauge und Leiter	33
* Durchbrüche in	32
- Isolationsabstand	27
- Kontrolle, der Registrierung	27
- Registrierung	27
Ionische Verunreinigung	
- unter Fotopolymermasken	53
- unter Siebdruckmasken	49
Isolationsabstand	
- auf Innenebenen	27
- Kohle/Kohle	58
- Kohle/Kupfer	58
- Mindestwert	22
- Reduzierung des	20
- um metallisierte Bohrungen herum	34
Isolationswiderstand, von Kohleelementen	63

K

Kantenschärfe	
- des Leiterbildes	23
- vom Kohledruck	58
- von abziehbaren Masken	68
Knospen, in metallisierten Bohrungen	39
Kohlefilm, auf Lötaugen	57
Kohlekontakte, Oberfläche	56
Kohleleiterbild, Detaillierungsgrad	57
Kohlepastendruck	
- abziehbare Masken, auf Kohleelementen	65
- Anschlußfläche	61
- Aushärtung, des Kohledrucks	66
- Delaminierung, von Kohleelementen	67
- Detaillierungsgrad, des Leiterbildes	57
- Dicke der Kupferfolie, zulässige	56
- Druckbreite, effektive	58
- Einbuchtungen, entlang Leiterkanten	59
- Fehlstellen, in der Kohleschicht	
* über Basismaterial	60
* über Kupfer	60
- Haftfestigkeit, des Drucks	63
- Isolationsabstand	
* Kohle/Kohle	58
* Kohle/Kupfer	58
- Isolationswiderstand, von Kohleelementen	63
- Kantenschärfe, des Leiterbildes	58
- Kohlefilm, auf Lötflächen	57
- Leiter, unter Kohleelementen	66
- Leiterbild	
* Anforderungen, Fünfundsiebzig-Prozent-Regel	58
* Registrierung	63

Stichwortverzeichnis

- Lösungsmittel, Beständigkeit gegen ... 57
- Löten, Auswirkung auf Kohleelemente ... 67
- Materialien ... 57
- Netzbildung, auf Kohleelementen ... 67
- Oberfläche, von Kontakten ... 56
- Oberflächenschutz, von Kohleelementen ... 65
- Prüfung, des Isolationswiderstandes
  * zwischen Kohleelementen ... 64
  * zwischen Kohlestreifen und Kupfer ... 64
- Prüfung, des Oberflächenwiderstandes, von Kohleelementen ... 64
- Prüfung, des Widerstandes, von Kohleelementen ... 64
- Qualitätsanforderungen ... 56
- Referenzmarken ... 63
- Risse, in Kohleelementen ... 67
- Rückstände, von Kohle ... 60
- Substrate, für den Kohledruck ... 66
- Übergang, Kohleelement/Anschlußfläche ... 66
- Überlappung
  * Kohle/Kupfer ... 61
  * Kohle/Lötstoppmaske ... 61
- Umgebungseinflüsse, auf Kohleelemente ... 67
- Unterschicht, für den Kohledruck ... 66
- Verträglichkeit, Kohle/abziehbare Maske ... 65
- Vorsprünge, entlang Leiterkanten ... 59

Komponentendruck
- Ausbluten ... 48
- Verwischen ... 48

Kratzer, im Leiterbild ... 15

Kupfer
- blankes, unter Siebdruckmasken ... 48
- blank/oxidiert, unter Fotopolymermasken ... 53
- unter Kohledruck ... 61

Kupferfolie
- Abheben ... 25
- Dicke ... 8
- Dicke, Festlegung ... 21

Kupfer, galvanisch aufgetragen
- des Leiterbildes ... 16
- in Bohrungen ... 15
  * Messung ... 15

Kurvenradius, von innenliegenden Ecken ... 79

L

Lackierung ... 71

Laminat
- Laminattypen ... 8
- Qualitätsanforderungen ... 11

Leiter
- Breite, minimale ... 22
- Reduzierung, der Breite ... 20
- unter Kohleelementen ... 66

Leiterbild
- Abheben des ... 26
- Änderung des ... 21
- allgemeine Anforderungen, Fünfundsiebzig-Prozent-Regel ... 20
- Dokumentation ... 18
- Einbuchtungen, entlang der Leiterkanten ... 24
- Fehlstellen ... 25
- Gleichmäßigkeit ... 19
- Haftfestigkeit ... 25
- Kantenschärfe ... 23
- Kohledruck, Detaillierungsgrad ... 57
- Metallpartikel ... 25
- Nadellöcher ... 25
- Position des, auf SMT-Leiterplatten ... 26
- Versatz
  * Messung des ... 22
  * von Kanten ... 22
- Vorsprünge, an den Leiterkanten ... 24

Leiterbild, Anforderungen für den Kohledruck ... 58

Leiterbildregistrierung, für den Kohledruck ... 63

Leiterplatten, fertig bearbeitete
- Anordnung, der Lagen ... 10
- Aufbau ... 10
- Basismaterial, starr ... 10
- Dicke ... 8
- Dickentoleranz ... 8
- dielektrisches Material ... 10
- Kennzeichnung, der Lagen ... 9
- Messung, der Schichtdicke ... 8
- mit Masseebenen ... 9
- mit Spannungsebenen ... 9
- Prepreglagen ... 10
- Reihenfolge, der Lagen ... 9

Leiterplatten, steckbare, Dicke von ... 8

Leiterplattenumriß, Bemaßung ... 78

Lochposition
- Festlegung der,
  * durch Bohrungsmarkierungen, auf dem Film ... 87
  * durch das Bohrprogramm ... 87
  * durch Koordinaten ... 87
- Toleranzen ... 87

Lochwand, von metallisierten Bohrungen
- Kupferdicke ... 15
- Porosität ... 38
- Unebenheit ... 38

Stichwortverzeichnis

Lösung der Hülse von der Lochwand 40
Lösungsmittel, Beständigkeit gegen
- von abziehbaren Masken 68
- von Fotopolymermasken 53
- von Kohledruck 57
- von Siebdruckmasken 48

Lötaugen bzw. -flächen
- Lochmontagetechnik,
  Reduzierung des Durchmessers 20
- Oberflächenmontagetechnik,
  Reduzierung der Abmessungen 20

Lötbarkeit, s. "Löten"
Lötbohrungen, Beschichtung der Lochkante 18

Löteffekte
- Temperatur 72
- Zeit 72

Löten
- Ausbläser, in Lötstellen 72
- Blasenbildung, im Basismaterial 72
- Delaminierung, im Basismaterial 72
- Lötbarkeit, Lagerbedingungen
  * Feuchtigkeit, relative 73
  * Temperatur 73
  * Verpackung 73
- Lötbarkeit, nach Lagerung
  * chemische Verzinnung 72
  * Heißluftverzinnung 73
  * Lackierung 73
  * Zinn/Bleigalvanisierung und Aufschmelzen 73
- Lötbarkeitsprüfung 71
- Löten, Auswirkungen 72
- Löttemperatur 72
- Lötzeit 71, 72
- Netzbildung 72
- Qualitätsanforderungen 71
- Risse, in metallisierten Bohrungen 72
- Tempern, der Leiterplatte 72

Löten, Auswirkungen auf Kohleelemente 67
Löt-/Entlötbeständigkeit, von metallisierten Bohrungen 35

Lötfehler
- Ausbläser, in Lötverbindungen 72
- Ausgasungen 72
- Blasenbildung, im Basismaterial 72
- Delaminierung, im Basismaterial 72
- Risse, in metallisierten Bohrungen 72

Löttemperatur 72
Lötzeit 71, 72

Lufteinschlüsse
- unter Fotopolymermasken 54
- unter Siebdruckmasken 49

M

Markierung, von Lagen 9
Maskendicke
- von Fotopolymer-Flüssigfilm 55
- von Fotopolymer-Trockenfilm 54
- von Siebdruckmasken 49

Masseebene
- in der fertig bearbeiteten Leiterplatte 9
- Zurücksetzen, entlang der Kanten 9

Materialien
- für den Kohledruck 57
- für Fotopolymermasken 54
- für Siebdruckmasken 49

Mechanische Bearbeitung
- Delaminierung
  * an den Leiterplattenkanten 73
  * nach der Bearbeitung 73
- des Umrisses 83
- Gewebezerrüttung (Crazing),
  nach der Bearbeitung 73
- Hofbildung (Haloing)
  * nach dem Stanzen 73
  * nach der Bearbeitung 73
- Qualitätsanforderungen 73

Metallisierte Bohrungen
- Durchmessertoleranz,
  * engere 31
  * normal 31
- Fehlstellen, in der Lochwand 35
- Galvanische Verkupferung 15
- Glasfasern, in der Lochwand 30
- Löt-/Entlötbeständigkeit 35
- Qualitätsanforderungen 30
- Restringe
  * Breite von 33
  * Unterbrechung, Anzahl 33
  * Unterbrechung, Isolationsabstand 34
  * Unterbrechung, Lage der 34
- Risse, umlaufende 35
- Sauberkeit, von Löchern 31
- Zinn/Blei-Galvanisierung 16

Metallisierte Bohrungen, Fehlerarten
- Ablösung der Hülse von der Lochwand 40
- chemische und galvanische Beschichtung,
  Verbindung zwischen 41
- Delaminierung, des Basismaterials 43
- Epoxidverschmierung (Smear) 36
- Glasfasern 30
- Gratbildung 40

Stichwortverzeichnis

- Harzrückzug 40
- Knospen 39
- Lochwand
  * porös 38
  * uneben 38
- Nagelkopfbildung 39
- Restringunterbrechung 34
- Risse 41
- Rückätzen, des Basismaterials 37
- Rückätzen, des Kupfers 37
- Taschen im galvanischen Niederschlag 39

Metallpartikel
- auf dem Leiterbild 25
- auf der Oberfläche, von Basismaterial 15, 25
- auf oder unter Fotopolymermasken 53
- auf oder unter Siebdruckmasken 48
- eingeschlossen, im Basismaterial 15

N

Nadellöcher
- im galvanischen Niederschlag, auf Randsteckkontakten 46
- im Leiterbild 25
- in Siebdruckmasken 49

Nagelkopfbildung, in metallisierten Bohrungen 39

Netzbildung
- aufgrund unvollständiger Aushärtung 14
- auf Kohleelementen 67
- auf Siebdruckmasken 49
- nach dem Löten 72

Nichtbenetzung, nach dem Aufschmelzen 17

Nichtmetallisierte Bohrungen
- Abheben, von Lotaugen 43
- Bedeckung, der Lochkante 44
- Durchmessertoleranz 44
- Grat, an der Lochkante 44
- Qualitätsanforderungen 43
- Restringe
  * Breite 44
  * Unterbrechung 44

Nickel-Zwischenschicht, für die galvanische Vergoldung 16

O

Oberfläche, von Kohlekontakten 56
Oberflächenfehler, aufgrund von Zinn/Bleibeschichtung 17
Oberflächenschutz
- chemische Verzinnung 71
- Heißluftverzinnung (Hot-Air-Levelling) 70
- Lackieren 71
- Qualitätsanforderungen 70
- Wassertauchlack 71
- Zinn/Bleigalvanisierung und Umschmelzen 70

Oberflächenschutz, von Kohleelementen 65

P

Plating-Pockets, in metallisierten Bohrungen 38
Porosität
- der Lochwand, von metallisierten Bohrungen 38
- des galvanischen Niederschlags, auf Randsteckkontakten 47

Prepregs
- Dicke 10
- Typen 8

Prüfung
- der fertigen Leiterplatte 30
- des Isolationswiderstandes
  * zwischen Kohleelementen 64
  * zwischen Kohlestreifen und Kupfer 64
- des Oberflächenwiderstandes, von Kohleelementen 64
- des Widerstandes, von Kohleelementen 64

Q

Qualitätsanforderungen
- an abziehbare Masken 67
- an das Aufschmelzen 17
- an das Basismaterial 11
- an das Laminat 11
- an das Leiterbild 19
- an den Kohledruck 56
- an den Oberflächenschutz 70
- an die galvanische Beschichtung 15
- an die Lötbarkeit 71
- an die mechanische Bearbeitung 73
- an Fotopolymermasken 53
- an metallisierte Bohrungen 30
- an nichtmetallisierte Bohrungen 43
- an Randkontaktleisten 45
- an Siebdruckmasken 48

R

Randkontaktleisten
- Ablösen, des galvanischen Niederschlages 45
- Anfasen 85
- Bedeckung, von Seitenkanten 45
- Codierschlitze

Stichwortverzeichnis

* Anordnung	84
* mechanische Herstellung	85
- Galvanisierung	
* Material	16
* Schichtdicke	16
- Haftfestigkeit, der galvanischen Schicht	47
- Nadellöcher, im galvanischen Niederschlag	46
- Porosität, des galvanischen Niederschlags	47
- Qualitätsanforderungen	45
- Registrierung von Ober- zu Unterseite	47
- Übergang, zwischen Kontakten und Leitern	46
Referenzbohrungen	
- Ausführung	77
- für im Nutzen montierte Leiterplatten	80
Referenzbohrungen im Nutzen	80
Referenzmarken, für den Kohledruck	63
Referenzsystem, Definition	77
Registriermarken, für die automatische Bestückung	27
Registrierung	
- von Innenlagen	27
* Kontrolle	27
Reihenfolge, der Lagen	9
Restringe, von metallisierten Bohrungen	
- Breite der Unterbrechungen	31
- Unterbrechung	
* Anzahl der Unterbrechungen	33
* Isolationsabstand	34
* Lage der Unterbrechung	34
Restringe, von nichtmetallisierten Bohrungen	
- Breite der Unterbrechung	44
- Unterbrechung	44
Risse	
- in Fotopolymermasken	53
- in Kohleelementen	67
- in metallisierten Bohrungen	41
* umlaufende Risse	35
Ritzen	83
Rückätzen (etch back)	
- von Basismaterial	37
- von Kupfer	37
Rückstände	
- vom Kohlepastendruck	60
- von Fotopolymermasken	53

S

Sauberkeit, von metallisierten Bohrungen	31
Schichtdicke des galvanischen Niederschlages, auf Steckkontakten	16
Schmelzen, von Zinn/Blei	17
Siebdrucklackrückstände, auf Lötaugen	48
Siebgedruckte Masken/Komponentenbezeichnungen	
- Abdecken, von Leitern	51
- Anordnung, des Komponentendrucks	50
- Aufbringung	47
- Ausbluten, des Drucks	48
- Ausfüllen, von Zwischenräumen zwischen Leitern	52
- Aushärten, der Maske	49
- blankes Kupfer, unter der Maske	48
- Delaminierung, unter der Maske	49
- Dicke, der Maske	49
- Ionische Verunreinigungen, unter Maske	49
- Lösungsmittel, Beständigkeit gegen	48
- Lufteinschlüsse, unter der Maske	49
- Materialien	49
- Metallpartikel, auf oder unter der Maske	48
- Nadellöcher, in der Maske	49
- Netzbildung, auf der Maske	49
- Qualitätsanforderungen	48
- Siebdruckrückstände, auf Lötaugen	48
- Überdrucken, von Lötflächen (Pads)	50
- Verwischen, des Drucks	48
- Zinn/Blei, unter der Maske	49
SMT-Leiterplatten	
- Automatische Bestückung	28
- Lage des Leiterbildes	26
Spannungsebenen	
- in der fertigen Leiterplatte	9
- Zurücksetzen, entlang den Kanten	9
Substrat, für den Kohledruck	66

T

Tempern, von Leiterplatten, vor dem Löten	73
Tenting, von Durchgangsbohrungen, mit Fotopolymermasken	53
Testabschnitt, Ausführung des	36
Trennung, der Hülse von der Lochwand	40

U

Überdrucken	
- von Lötflächen, mit Siebdruckmasken	50
Übergang	
- Kohleelement/Anschlußfläche	66
- Leiter/Randkontaktleiste	46
Überlappung	
- Kohle/Kupfer	61
- Kohle/Lötstoppmaske	61
- von Lötaugen bzw. -flächen	
* mit Fotopolymermasken	55

Stichwortverzeichnis

* mit Siebdruckmasken	50
Umgebungsanforderungen, an Kohleelemente	67
Umriß, der Leiterplatte	
- Bemaßung	78
- Festlegung	
* durch Bemaßung	82
* durch Eckenmarkierungen auf dem Film	82
* durch Fräsprogrammdaten	82
- mechanische Bearbeitung	83
Unebenheit	
- des Komponentendrucks	48
- des Leiterbildes	23
- von Siebdruckmasken	48
Unterschicht, für den Kohledruck	66

V

Verbindung von chemischem und galvanischem Kupfer, in metallisierten Bohrungen	41
"Verbrennen", beim Galvanisieren	15
Verfärbung, des Basismaterials, nach dem Aufschmelzen	17
Vergoldung, galvanische	
- Anforderungen, an Goldschichten	17
- mit Nickelunterschicht	16
- Schichtdicke	17
Verschließen, von Bohrungen, mit abziehbaren Masken	69
Verträglichkeit, Kohle/abziehbare Maske	65
Verwindung	
- Anforderungen,	
* normal	75
* schärfere	76
- Bestimmung der	74
- Definition der	73
- Normalverteilung	76
Vorsprünge, entlang Leiterkanten	24
- beim Kohledruck	59

W

Wareneingangskontrolle, der Fotowerkzeuge	19
Wassertauchlack	71
Widerstand, von Kohleelementen	63
Wölbung	
- Anforderungen	
* normal	75
* schärfer	76
- Definition	73
- Messung der	74
- Normalverteilung	76

Z

Zentriermarken, optische, für die automatische Bestückung	29
Zinn/Blei	
- unter Fotopolymermasken	53
- unter Siebdruckmasken	48
Zinn/Blei-Galvanisierung	16
- Aufschmelzen	70
- Erstarrungslinien, nach dem Aufschmelzen	17
- Legierungszusammensetzung	16
- Oberfläche	17
- Oberflächenfehler	17
Zinn/Bleiüberhang, Entfernung des	17
Zinngehalt, Streubreite der Legierungszusammensetzung	16

Index

A

Additive techniques ... 9
Adhesion, of
- carbon printing ... 63
- conductors ... 25
- peelable masks ... 67
- plating
  * of edge connector contacts ... 47
  * of pattern ... 15
- SMT solder pads ... 26

Air pockets
- under photopolymer masks ... 54
- under screen-printed masks ... 49

Annular rings, of nonplated-through holes
- break-out ... 44
- width of ... 44

Annular rings, of plated-through holes
- break-out
  * insulation distance ... 34
  * location of ... 34
  * number of ... 33
- width of ... 31

AOI, automatic optical inspection ... 30
Apertures, in inner layer planes ... 31
Automatic assembly
- edge clearance area ... 28
- fiducials ... 29
- SMT boards ... 28
- targets, optical ... 29
- tooling holes ... 29

Automatic optical inspection ... 30

B

Baking, of boards, before soldering ... 73
Balancing, of pattern ... 19
Base laminate, rigid ... 10
Base material
- blistering in ... 12
- curing of ... 14
- delamination of ... 12
- haloing in ... 13
- metallic inclusions ... 15
- swelling of ... 12
- types of ... 8
- with weave exposure ... 14
- with weave texture ... 14

Bleeding, of screen-printed masks ... 48
Blistering, in base material ... 12

Board contour, dimensioning of ... 78
Boards, finished
- base laminate, rigid ... 10
- build-up of ... 10
- dielectric material ... 10
- distribution, of layers ... 10
- marking, of layers ... 9
- measurement, of thickness ... 8
- prepreg layers ... 10
- sequence, of layers ... 9
- thickness of ... 8
- thickness tolerance of ... 8
- with ground planes ... 9
- with voltage planes ... 9

Boards, plugable, thickness of ... 8
Build-up, of finished board ... 10
Burning, of plating ... 15
Burrs
- at nonplated-through holes ... 44
- in plated-through holes ... 40

C

CAD plotter data ... 18
Carbon contacts, surface of ... 56
Carbon film, on solder pads ... 57
Carbon pattern, degree of detail ... 57
Carbon printing
- adhesion, of printing ... 63
- carbon film, on solder pads ... 57
- compatibility, carbon/peelable mask ... 65
- conductors, under carbon elements ... 66
- cracks, in carbon elements ... 67
- curing, of carbon printing ... 66
- degree of detail, of pattern ... 57
- delamination, of carbon elements ... 67
- edge definition, of pattern ... 58
- environmtl. requiremts., of carbon elements ... 67
- indentations, along pattern edges ... 59
- insulation distance
  * carbon/carbon ... 58
  * carbon/copper ... 58
- insulation resistance, of carbon elements ... 63
- interface, carbon element/termination area ... 66
- materials ... 57
- overlapping
  * carbon/copper ... 61
  * carbon/solder mask ... 61
- pattern
  * registration ... 63

Index

- * requirements, 75 % rule 58
- peelable masks, on carbon elements 65
- printing width, effective 58
- projections, along pattern edges 59
- quality requirements 56
- reference marks 63
- remnants, of carbon 60
- solvents, resistance towards 57
- soldering effects, on carbon elements 67
- substrate, for carbon printing 66
- surface, of contacts 56
- surface protection, of carbon elements 65
- termination area 61
- test, of insulation resistance
  - * between carbon elements 64
  - * between strap and copper 64
- test, of resistance, of carbon elements 64
- test, of surface resistance, of carbon elements ... 64
- thickness, of copper foil, acceptable 56
- underlay, for carbon printing 66
- voids, in carbon layer
  - * over copper 60
  - * over base material 60
- webbing, on carbon elements 67

Chamfering, of edge connector 85
Cleanliness, of plated-through holes 31
Clogging, of holes, with peelable mask 69
Compatibility, carbon/peelable mask 65
Component notation
- bleeding 48
- blur 48

Conductors
- reduction, of width 20
- under carbon elements 66
- width, minimum 22

Contour, of board
- determination of
  - * by dimensioning 82
  - * by router data 82
  - * by corner marks, on film 82
- machining of 83
- dimensioning of 78

Copper
- bare, under screen-printed masks 48
- bare/oxidized, under photopolymer masks 53
- under carbon printing 61

Copper foil
- lifting of 25
- thickness of 8
- thickness, selection of 21

Copper plating
- in holes 15
  - * measurement of 15
- of pattern 16

Corners, inward, radius of curvature 79

Coverage
- of conductors 56
  - * with photopolymer masks 56
  - * with screen-printed masks 51
- of pattern, with peelable masks 69
- of side edges, of edge connector contacts 45
- of solder pads, with peelable mask 69

Cracks
- in carbon elements 67
- in photopolymer masks 53
- in plated-through holes 41
  - * circumferential cracks 35

Crazing
- after machining 73
- in base material 12

Curing
- of carbon printing 66
- of photopolymer masks 54
- of screen-printed masks 49

D

Degree of detail, of carbon pattern 57
Delamination
- after machining 73
- along board edges 73
- of base material 13
  - * locally 13
- in base material 43
- of carbon elements 67
- under photopolymer masks 54
- under screen-printed masks 49

Dewetting, after reflowing 17
Diameter tolerance
- of nonplated-through holes 44
- of plated-through holes 31

Dielectric material
- thickness of 10
- thickness tolerance of 11

Dimensional changes, of pattern, permanent 26
Dimensioning
- of board contour 78
- of panelized boards 80

Discolouration, of base material, after reflowing ... 17
Distribution, of layers 10
Documentation, assessment of 21

Index

E

Edge clearance area, for automatic assembly 28
Edge connectors
- adhesion, of plating 47
- chamfering of 85
- coverage, of side edges 45
- flaking, of plating 45
- interface, between contacts and conductors 46
- pinholes, in plating 46
- plating
  * material 16
  * thickness 16
- polarization slot
  * location of 84
  * machining of 85
- porosity, of plating 47
- quality requirements 45
- side-to-side registration 47

Edge definition
- of carbon printing 58
- of pattern 23
- of peelable masks 68

Electrical tests 30
Environmtl. requiremts., of carbon elements 67
Epoxy smear, in plated-through holes 36
Etch factor, requirements of 22
Etchback
- of base material 37
- of copper 37
Execution, of peelable masks 67

F

Fiducials, for automatic assembly 27
Filling, of conductor spaces
- with photopolymer dry-film mask 54
- with photopolymer liquid-film mask 55
- with screen-printed mask 51
Films
- derived from CAD plotter data 18
- photographically reduced 18
Fixing tabs, for panelized boards 80
Flaking, of plating, of edge connector contacts 45
Freeze lines, in tin/lead layer, after reflowing 17
FR4 8

G

Glass fibres, in plated-through holes 30
Gold plating 16

- requirements, of gold 17
- thickness of layer 17
- with nickel underlay 16
Ground plane
- in finished board 9
- withdrawal of, along edges 9

H

Haloing
- after machining 73
- after punching 73
- in base material 13
Heat effect, on peelable masks 68
Hole edge coverage, nonplated-through holes 44
Hole position
- determination of
  * by drill tape 87
  * by coordinates 87
  * by hole markings, on film 87
- tolerances of 87
Hole wall, of plated-through holes
- copper thickness 15
- porosity 38
- unevenness 38
Hole wall pull-away, in plated-through holes 40
Holes, types of 86

I

Immersion tin 71
Incoming inspection, of film set 19
Indentations
- along pattern edges 24
- of carbon printing 59
Ink film, on solder pads, from screen-printing 48
Inner layers
- checking, of registration 27
- insulation distance 27
- planes
  * apertures in 32
  * distance, between solder pad and conductor 33
- registration of 27
Insulation distance
- around plated-through holes 34
- carbon/carbon 58
- carbon/copper 58
- minimum value 22
- on inner layers 27
- reduction of 20

Index

Insulation resistance, of carbon elements 63
Interface
- carbon element/termination area 66
- conductor/edge connector contact 46
Ion-contamination
- under photopolymer masks 53
- under screen-printed masks 49

L

Lacquering 71
Laminate
- quality requirements of 11
- types of 8
Laminate/hole wall separation 40
Lifting, of copper 26, 43
Location, of component notation 50

M

Machining
- crazing, after machining 73
- delamination
  * after machining 73
  * along board edges 73
- haloing
  * after machining 73
  * after punching 73
- of contour 83
- quality requirements 73
Machining tolerances
- when punching 83
- when routing 83
- when sawing 83
- when scoring 83
- when shearing 83
Marking, of layers 9
Mask thickness
- of photopolymer dry-film 54
- of photopolymer liquid-film 55
- of screen-printed masks 49
Materials
- for carbon printing 57
- for photopolymer masks 54
- for screen-printed masks 49
Measling, in base material 11
Melting, of tin/lead 17
Metal particles
- encapsulated, in base material 15
- on pattern 25
- on or under photopolymer masks 53
- on or under screen-printed masks 48
- on surface, of base material 15, 25

N

Nailheading, in plated-through holes 39
Nickel underlay, for gold plating 16
Nodules, in plated-through holes 39
Nonplated-through holes
- annular rings,
  * break-out 44
  * width of 44
- burrs, along hole edge 44
- coverage, of hole edge 44
- diameter tolerance 44
- lifting, of solder pads 43
- quality requirements 43
Nonwetting, after reflowing 17

O

Overlapping
- carbon/copper 61
- carbon/solder mask 61
- of solder pads
  * with photopolymer masks 55
  * with screen-printed masks 50
Overprinting
- of solder pads, with screen-printed masks 50

P

Panel reference holes 80
Panelized boards
- dimensioning of 80
- distance between 80
- fixing tabs 80
Pattern
- adhesion of 25
- balancing of 19
- carbon, degree of detail 57
- change of 21
- displacement
  * measurement of 22
  * of edges 22
- documentation of 18
- edge definition of 23
- general requirements, 75 % rule 20
- indentations, along pattern edges 24
- lifting of 26
- metal particles in 25

Index

- pinholes in 25
- position of, on SMT boards 26
- projections, along pattern edges 24
- voids in 25

Pattern registration, for carbon printing 63
Pattern requirements, of carbon printing 58
Peelable masks
- adhesion of 67
- application of 67
- closing, of holes 69
- coverage
  * of pattern 69
  * of solder pads, on opposite side 69
- edge definition 68
- execution of 67
- heat effects 68
- on carbon elements 65
- quality requirements 67
- removal, of mask 68
- solvents, resistance towards 68
- voids, unintentional 70

Photopolymer masks
- air pockets, under mask 54
- application 52
- copper, bare/oxidized, under mask 53
- cracks, in mask 53
- curing, of mask 54
- delamination, under mask 54
- ion-contamination, under mask 53
- materials 54
- metal particles, on or under mask 53
- quality requirements 53
- remnants, of mask 53
- removal, of masks 54
- solvents, resistance towards 53
- tenting, of via holes 53
- tin/lead, under mask 53
- unintentional 53

Phototools, with compensation 19
Pinholes
- in pattern 25
- in plating, of edge connector contacts ... 46
- in screen-printed masks 49

Plated-through holes
- annular rings
  * break-out, insulation distance 34
  * break-out, location of 34
  * break-out, number of 33
  * width of 33
- cleanliness, of holes 31

- copper plating 15
- cracks, circumferential 35
- diameter tolerance
  * ordinary 31
  * tighter 31
- glass fibres, in hole wall 30
- quality requirements of 30
- soldering/unsoldering strength 35
- tin/lead plating 16
- voids, in hole wall 35

Plated-through holes, defect types
- annular ring break-out 34
- burrs 40
- cracks 41
- delamination, of base material 43
- epoxy smear 36
- etchback, of base material 37
- etchback, of copper 37
- glass fibres 30
- hole wall
  * porous 38
  * uneven 38
- hole wall pull-away 40
- laminate/hole wall separation 40
- nailheading 39
- nodules 39
- plating contact 41
- plating pockets 39
- resin recession 40

Plating
- adhesion of 15
- balancing, of pattern 20
- burning of 15
- quality requirements of 15
- with copper 15
- with gold 17
  * on nickel 16
- with tin/lead 16
  * composition of alloy 16

Plating contact, in plated-through holes 41
Plating materials, for edge connector contacts .. 16
Plating pockets, in plated-through holes 38
Plating thickness, of edge connector contacts .. 16
Polarization slot, in edge connector
- location of 84
- machining of 85
- position of 84

Porosity
- of hole wall, of plated-through holes 38
- of plating, of edge connector contacts .. 47

Index

Prepregs
- thickness 10
- types of 8

Printing width, of carbon printing 58
Production films, with compensation 19
Projections
- along pattern edges 24
- of carbon printing 59

Q

Quality requirements
- of base material 11
- of carbon printing 56
- of edge connector contacts 45
- of laminate 11
- of machining 73
- of nonplated-through holes 43
- of pattern 19
- of peelable masks 67
- of photopolymer masks 53
- of plated-through holes 30
- of plating 15
- of reflowing 17
- of screen-printed masks 48
- of solderability 71
- of surface protection 70

R

Radius, of curvature, of inward corners 79
Reference holes
- execution of 77
- for panelized boards 80

Reference marks, for carbon printing 63
Reference system, definition of 77
Reflowing
- coverage
  * of conductor edges 17
  * of hole-to-pad interface 18
- discolouration, of base material 17
- execution of 18
- incomplete 17
- of tin/lead 17
- quality requirements, of surface 17
- with dewetting 17
- with nonwetting 17

Registration
- of inner layers 27
  * checking of 27

Remnants
- of carbon printing 60
- of photopolymer masks 53

Removal
- of peelable masks 68
- of photopolymer masks 54

Resin recession, in plated-through holes 40
Resistance, of carbon elements 63

S

Scoring 83
Scratches, of pattern 15
Screen-printed masks/component notations
- air pockets, under mask 49
- application of 47
- bare copper, under mask 48
- bleeding, of print 48
- blur, of print 48
- coverage, of conductors 51
- curing, of mask 49
- delamination, under mask 49
- filling, of conductor spaces 52
- ink film, on solder pads 48
- ion-contamination, under mask 49
- materials 49
- metal particles, on or under mask 48
- overprinting, of solder pads 50
- pinholes, in mask 49
- position, of component notation 50
- quality requirements 48
- solvents, resistance towards 48
- thickness, of mask 49
- tin/lead, under mask 48
- webbing, on mask 49

Separation, between laminate and hole wall ... 40
Sequence, of layers 9
Seventy-five % rule
- of carbon pattern 58
- of copper pattern 19

Side-to-side registration, of edge contacts 47
SMT boards
- automatic assembly 28
- pattern position 26

Solder coating and hot-air levelling 70
Solder holes, coverage, of hole-to-pad interface ... 18
Solder pads
- HMT, reduction, of diameter 20
- SMT, reduction, of dimensions 20

Solderability, see "Soldering"

Index

Soldering
- baking, preceding, of board 72
- blistering, in base laminate....................................72
- blowholes, in solder joints 72
- cracks, in plated-through holes 72
- delamination, in base material 72
- quality requirements...71
- solderability, after storage
  * immersion tin..72
  * lacquering 73
  * solder coating and hot-air levelling....................73
  * tin/lead plating and reflowing...........................73
- solderability, storage conditions
  * temperature 73
  * humidity, relative 73
  * packing 73
- solderability test 71
- soldering effect 72
- soldering temperature 72
- soldering time..71, 72
- webbing...72

Soldering defects
- blistering, in base material 72
- blowholes, in solder joints 72
- cracks, in plated-through holes 72
- delamination, in base material 72

Soldering effect
- temperature 72
- time 72

Soldering effects, on carbon elements 67
Soldering temperature 72
Soldering time..71, 72
Soldering/unsoldering strength, PTH 35
Solvents, resistance towards,
- of carbon printing 57
- of peelable masks 68
- of photopolymer masks 53
- of screen-printed masks 48

Substrate, for carbon printing...................................66
Surface, of carbon contacts 56
Surface defects, due to tin/lead plating......................17
Surface protection
- immersion tin..71
- lacquering 71
- quality requirements..70
- solder coating and hot-air levelling.........................70
- tin/lead plating and reflowing 70
- water-dip lacquer 71

Surface protection, of carbon elements 65

Swelling, local, of base material................................12

T

Tabs, fixing, for panelized boards 80
Targets, optical, for automatic assembly....................29
Tenting, of via holes, with photopolymer masks 53
Termination area, for carbon printing.........................61
Test
- of finished board 30
- of insulation resistance
  * between carbon elements..............................64
  * between strap and copper 64
- of resistance, of carbon elements 64
- of surface resistance, of carbon elements 64

Test coupon, execution of..36
Thickness
- of copper
  * in plated-through holes 15
  * on conductors 16
- of copper foil 8
  * for carbon printing...56
- of finished board 8
  * measurement of 8
- of gold 17
- of photopolymer dry-film masks 54
- of photopolymer liquid-film masks 55
- of plugable boards 8
- of screen-printed masks......................................49

Thickness tolerance, of finished board........................8
Tin content, variation, of alloy composition 16
Tin/lead
- under photopolymer masks...................................53
- under screen-printed masks..................................48

Tin/lead overhang, removal of 17
Tin/lead plating..16
- composition of alloy 16
- freeze lines, after reflowing 17
- reflowing of..70
- surface defects...17
- surface of..17

Tooling holes, for automatic assembly 29
Twist
- definition of..73
- determination of 74
- requirements,
  * normal 75
  * tighter 76
- normal distribution..76

- 103 -

Index

U

Underlay, for carbon printing 66
Unevenness
- of component notation 48
- of pattern 23
- of screen-printed masks 48

V

Voids
- in carbon printing
  * on base material 60
  * on copper 60
- in pattern 25
- in peelable masks 70
- in plated-through holes 35
Voltage planes
- in finished board 9
- withdrawal of, along edges 9

W

Warp
- definition of 73
- determination of 74
- requirements
  * normal 75
  * tighter 76
- normal distribution 76
Water-dip lacquer 71
Weave exposure, of base material 14
Weave texture, of base material 14
Webbing
- after soldering 72
- due to incomplete curing 14
- on carbon elements 67
- on screen-printed masks 49

Firmen-Verzeichnis
zum Anzeigenteil

Blasberg Oberflächentechnik GmbH,
    D-5650 Solingen 11 .... 5A

Bohncke & Gissel GmbH,
    D-6270 Idstein 4A

Galvanex GmbH,
    D-6074 Rödermark 2 .... 6A

Hehl Elektronik GmbH,
    D-5657 Haan 1 .... 7A

Ingo-Chemie GmbH,
    D-8000 München 60 .... 6A

Andreas Maier GmbH & Co.KG,
    D-7959 Schwendi-Hörenhausen 2A

Moderne elemat GmbH,
    D-7000 Stuttgart 23 .... 7A

Norplex OAK Europa GmbH,
    D-5272 Wipperfürth 2A

Pill, Anlagen zur Oberflächenbehandlung von Leiterplatten,
    D-7159 Auenwald 1A

Emil Schmitz (GmbH & Co.),
    D-5650 Solingen 1 .... 3A

Sondermann Pumpen + Filter GmbH,
    D-5000 Köln 21 .... 8A